Banking on the Environment

Banking on the Environment

Multilateral Development Banks and Their
Environmental Performance in Central and
Eastern Europe

Tamar L. Gutner

The MIT Press
Cambridge, Massachusetts
London, England

This book was set in Sabon by Achorn Graphic Services, Inc., with the Miles 33 system and printed and bound in the United States of America.

Library of Congress Cataloging-in-Publication Data

Gutner, Tamar L.
  Banking on the environment : multilateral development banks and their environmental performance in Central and Eastern Europe / Tamar L. Gutner.
    p. cm. — (Global environmental accord)
Includes bibliographical references (p.   ) and index.
ISBN 0-262-07235-1 (hc. : alk. paper) — ISBN 0-262-57159-5 (pbk. : alk. paper)
  1. Development banks—Europe, Central. 2. Development banks—Europe, Eastern. 3. Environmental policy—Europe, Central. 4. Environmental policy—Europe, Eastern. 5. Economic development—Environmental aspects—Europe, Central. 6. Economic development—Environmental aspects—Europe, Eastern. I. Title. II. Global environmental accord.

HG1976.C36 G88   2002
333.7′0943—dc21
                                                        2002021917

Contents

Series Foreword

A new recognition of profound interconnections between social and natural systems is challenging conventional constructs and the policy predispositions informed by them. Our current intellectual challenge is to develop the analytical and theoretical underpinnings of an understanding of the relationship between the social and the natural systems. Our policy challenge is to identify and implement effective decision-making approaches to managing the global environment.

The series on Global Environmental Accord adopts an integrated perspective on national, international, cross-border, and cross-jurisdictional problems, priorities, and purposes. It examines the sources and the consequences of social transactions as these relate to environmental conditions and concerns. Our goal is to make a contribution to both intellectual and policy endeavors.

Nazli Choucri

Acknowledgments

This book began as a dissertation prospectus in the Political Science Department at MIT. At the time, I was struck by the World Bank's efforts to fashion itself a leading actor in global environmental governance amid unrelenting criticism from nongovernmental organizations. As a former financial journalist who covered banking and development finance issues for AP–Dow Jones in New York and London, I was curious about what being "green" meant in a large international financial institution. What was the explanation for the seeming gap between the Bank's rhetoric and its alleged actions?

I decided a comparative study was necessary to understand the factors that shape the ability of a multilateral development bank (MDB) to respond to environmental mandates. I also felt it was important to focus on the activities of a set of these institutions in a common geographical region. Central and Eastern Europe was the obvious choice, because it most challenged an MDB's ability to harmonize its economic development and environmental protection and management strategies. The World Bank and two other MDBs—the European Bank for Reconstruction and Development (EBRD) and European Investment Bank (EIB)—were among the top donors in the region. As I developed the topic, I was surprised to learn that virtually no research existed on the environmental behavior of the EIB. This was curious, given the level of attention and criticism that surrounds the World Bank's actions, while the EIB now lends more each year than the World Bank but has made far fewer efforts to reform itself in ways that increase its performance and transparency. The EIB is a global actor, even though the majority of its lending is to European Union member states.

Since this project began, the World Bank and its sister institutions, the International Monetary Fund and the World Trade Organization, have become the targets of increasingly prominent and sometimes violent antiglobalization protests. These demonstrations intensified ongoing discussions and debate about whether and how these institutions should be reformed. It is also only a matter of time before the antiglobalization protestors turn their ire toward the EIB. While this book analyzes the environmental performance of the World Bank, EBRD, and EIB with a focus on their environmental behavior in Central and Eastern Europe, my goal is to contribute to scholarly and policy thinking on how different political and institutional factors influence the ability of an MDB to take on new mandates and implement them. International organizations have become simultaneously more necessary and more reviled. It is therefore critical for policymakers, scholars, and activists to better understand the sources of institutional inertia and innovation in thinking about how to improve institutional performance.

Many individuals and organizations deserve thanks for helping me realize this book. I benefited from insightful comments, not to mention encouragement, from Barbara Connolly, David Fairman, Dalia Dassa Kaye, Bob Keohane, and Inger Weibust. I am also grateful to M. J. Peterson and two anonymous reviewers for valuable substantive suggestions. My dissertation committee, Kenneth Oye, Lawrence Susskind, and Eugene Skolnikoff, ably guided me through the dissertation process. I owe a great debt to the many people I interviewed for this project; in fact, "interview" is too narrow a word to encompass all the kinds of assistance these individuals offered, which included reading and commenting on my dissertation and book manuscript, introducing me to others, providing me with key documents, and more. Many thanks to Richard Ackermann, Wendy Ayres, Luca Barbone, Dan Berg, Peter Carter, Judit Galambos, Kristalina Georgieva, Jozsef Feiler, Brendan Gillespie, Axel Hörhager, Bill Kennedy, Stephen Lintner, Eric Meyer, Gerry Muscat, Jean-Jacques Schul, Sari Söderström, Helmut Schreiber, Tomasz Terlecki, Josué Tanaka, and Linus Vainius.

For research assistance, I wish to thank Hamilton Bean, Maia Curtis, Leanne Johnson, Daniel Keegan, Vivi Mazarakis, and Kevin Shehan. I am especially grateful to the entire staff of the Brookings Institution library, who were extremely helpful during my year-long stay. Many thanks also to Clay Morgan and Judy Feldmann at MIT Press for their help and hard work.

I would like to acknowledge the invaluable financial support I received from the Social Science Research Council (SSRC), the Brookings Institution, the Academic Council on the United Nations System (ACUNS), the Institute for the Study of World Politics, MIT's Center for International Studies, the MIT-Volvo Award for Environmental Research, Harvard University's Center for European Studies, the American Council of Learned Societies/SSRC, the Harvard-MIT MacArthur Transnational Security program, and American University. MIT, Harvard, Brookings, and American also provided me with challenging intellectual environments that were so important to me throughout the process.

My deepest gratitude is to my family and friends for their love and support. My husband, Max Holland, has seen me through the end of this journey, with the right mix of encouragement and perspective.

This book is dedicated to my late grandfather, Aaron Gunther, who encouraged me from a young age to learn about the world around me and supported all my efforts to do so.

Central and Eastern Europe

1

Introduction and Overview

... the World Bank (is) at the centre of the movement towards a more sustainable way of life on our planet. ...

—Maurice Strong[1]

If it is true as Maurice Strong says, that the Bank is "at the center of the movement towards a more sustainable way of life on our planet" ... we are in deep trouble as a planet and a species. ... The Bank has failed to introduce policies which will implement sustainable development, principally because the development paradigm within which it operates has not changed.

—Lisa Jordan[2]

One of the striking features of global governance in the post–Cold War era is the increasing responsibility nation-states have given international organizations (IOs) to pursue collective solutions to global and regional problems. Shareholder states are asking IOs to take on a variety of new policy issues that did not exist when these institutions were created, and the ability of IOs to implement new mandates is critical in a period when patterns of global governance are becoming both more diffuse and complex. Perhaps the most dramatic illustration of this trend in the late twentieth century is NATO's shift from a collective defense organization designed to deter Soviet attack, to one involved in fighting inside non–member states, where the driving issues are ethnic conflict and human rights abuses. The United Nations, too, is struggling to address new demands for peacekeeping operations in areas where its mandate is not always clear, from Somalia to Kosovo. In international economic affairs, the International Monetary Fund (IMF) has become a central actor in "bailing out" countries with collapsing economies, while the World Trade Organization (WTO) has more powers to settle trade disputes among states than envisioned for its predecessor, the General Agreement on Tariffs and Trade (GATT).

This trend is also highly visible in the arena of global environmental politics, where rising international concern over transboundary environmental problems has resulted in major new responsibilities for IOs. Perhaps ironically, the most important set of IOs asked to take a leading role in financing environmental improvement is the one most criticized by nongovernmental organizations (NGOs) for poor environmental performance. These IOs are multilateral development banks (MDBs), international financial institutions that are struggling to find ways to connect their traditional mandates of furthering economic development and alleviating poverty with a number of new roles, of which promoting environmental protection has been among the most publicized. As shareholder countries look to MDBs for leadership in global and regional negotiations and actions on environmental issues, many people criticize these very institutions for doing a poor job following their own environmental policies, with sometimes disastrous results in terms of increasing environmental degradation in recipient countries. Environmental criticism is also contributing to larger movements against international financial and trade organizations by antiglobalization protesters whose campaigns have increased since the 1999 World Trade Organization meetings in Seattle.

This book investigates the environmental behavior of three major multilateral development banks, in order to determine whether or not these institutions are making great strides in integrating environmental considerations into their work, or whether they are engaged in "green" window dressing and causing more harm than good in recipient countries. The three banks under discussion are the World Bank, the European Bank for Reconstruction and Development (EBRD), and the European Investment Bank (EIB), and the book examines and compares the process by which each bank received and institutionalized new environmental policy mandates, and how these mandates were implemented in one common geographical region: Central and Eastern Europe (CEE). CEE is defined in this book as including Albania, Bosnia and Herzegovina, Bulgaria, Croatia, the Czech Republic, Estonia, Hungary, Latvia, Lithuania, FYR Macedonia, Poland, Romania, the Slovak Republic, and Slovenia. My goal is in part to explain what factors shape the environmental behavior and performance of a multilateral development bank, but probing the actions of these MDBs in this region allows other important questions be explored as well: first, how have these key donors fared in a region that most challenges their ability to harmonize their economic and environmental strategies? And second, why have three MDBs with similar sets of major shareholders adopted such different approaches to their environmental activities in the region?

The collapse of communism in CEE revealed the most polluted countries in Europe, and a region where both economic restructuring and environmental clean-up and reform needs were acute. These three banks, in turn, are among the largest donors in CEE, together agreeing to lend over $40 billion in the period fiscal 1990–2000. Environmental aid was an early priority of donor countries working in the transition countries, in order to help address the environmental damage throughout CEE, and to assist in environmental policy reform efforts. While environmental assistance was quickly overshadowed by the more overwhelming demands of economic restructuring, by the late-1990s attention to environmental policy issues gained fresh momentum as the new countries hoping to join the European Union (EU) began to tackle the enormous challenge of bringing their environmental standards up to European levels—a process the European Commission estimated would total a daunting $130 billion.[3]

An examination of the activities of these three banks in Central and Eastern Europe produces somewhat counterintuitive findings. The World Bank, the most environmentally scrutinized and criticized of the three, has responded most successfully to its environmental mandate in CEE, with the greatest share of "green" projects and the largest number of agenda-setting activities that have helped CEE governments develop new environmental policies and institutions. By the late 1990s, however, its leverage began to decline within countries with access to alternative sources of financing, with less conditionality attached. The EIB, by contrast, is the least well known of the three MDBs but also possesses the weakest environmental policies of the three, resulting in a minimal environmental response in CEE. The Luxembourg-based EIB was created by the 1957 Treaty of Rome to be the European Community's (now the EU's) long-term lending institution, but it is in fact a global actor that now lends more money each year than the World Bank.[4] While the vast majority of its lending goes to EU member states, the EIB operates in 120 countries outside of the EU and is poised to be the most important—and eventually, the *only*—MDB in CEE countries joining the EU. NGOs have only recently begun to scrutinize this secretive institution, while scholarly attention is virtually nonexistent. The London-based EBRD, in turn, was founded in 1990 with a stated commitment to addressing sustainable development issues in postcommunist countries. It maintains an intermediate position between the other two banks, more dynamic than the EIB in its environmental activities, but with less breadth than the World Bank.

Overall, the World Bank has played an important role in providing intellectual and policy support for environmental reform in the region, and its policy activities have been complemented by a number of projects with primary environmental goals or significant environmental components. The EBRD is less involved in policy work, and has defined its environmental projects far more narrowly—mainly as those undertaken by its municipal and energy efficiency units—although the volume of its environmental lending is approaching that of the World Bank. The EIB has undertaken very little policy work and has financed the fewest environmental projects. These projects mainly reflect recipient demand or the EIB's role as cofinancier of projects that one of the other two MDBs designed. In the period 1990–1999, which is the focus of this book, projects with primary environmental components or significant environmental objectives totaled approximately $1.3 billion for the World Bank, $1.1 billion for the EBRD, and $605 million for the EIB.[5] These trends continued in subsequent years, with the World Bank holding its position as most active in financing such projects. These figures, however, do not include financing for technical assistance, agenda-setting exercises, and other nonloan activities, which further reflect the greater depth and scope of the World Bank's activities, followed by the EBRD.

None of the banks has been involved in financing in the CEE region the types of enormously environmentally destructive projects that MDBs—particularly the World Bank—have been accused of financing elsewhere, projects that have destroyed coastal ecosystems, promoted enormous deforestation, exhausted soils, and displaced thousands of people.[6] At the same time, the environmental performance of all three banks has been far from the lofty goals expressed in their policies and public documents.

To different degrees, they have financed projects that are controversial from an environmental viewpoint, where the priorities or impact of projects appear to conflict with their stated policies, or where the policies themselves are poorly followed. Efforts to "green" the MDBs have faced a variety of obstacles, including diffuse governance structures, a mismatch between new environmental mandates and existing patterns of institutional design and incentives, and difficulty in "selling" environmental projects or conditionality to recipients. Even where all three banks have been involved in projects with significant environmental goals or components, they have faced numerous challenges in the implementation phase, as they grapple with domestic political, technical, and other issues. The World Bank, the most progressive of the three, admitted its own shortcomings in an April 2000 report by its Operations Evaluation Department, which noted that "the impact of World Bank Group programs on broad environmental trends in the developing world has been limited, and the achievements of various programs have been mixed."[7]

What factors influence the environmental behavior of these MDBs? My central argument is that the depth and scope of an MDB's environmental activity, and the differences between these three MDBs, reflect the interaction of two factors: the degree of shareholder commitment to environmental issues, and how demand-driven or "banklike" the MDB is designed to be. The book traces the ways in which external pressure from major shareholder countries, usually supported or pushed by environmental NGOs, has been a key factor determining the level of an MDB's commitment to addressing environmental issues, as seen in its stated policy goals. Shareholder commitment to environmental issues is relatively stronger at the World Bank, and virtually absent at the EIB.

However, shareholder commitment is necessary but not sufficient in explaining the banks' environmental behavior. Shareholders play an important role on a strategic level, but new environmental mandates do not always fit well with existing institutional design and incentive systems. As a result, even in cases where shareholders have expressed a strong desire for the MDB to address environmental issues, institutional design and incentive systems play a critical role in how these environmental objectives are translated into the banks' activities. The degree to which an MDB is demand-driven, and behaves more like a financial institution than a development agency, greatly affects its efforts to address environmental issues in its work and to sell environmental projects to recipient countries or actors.

All MDBs are designed in some ways to behave like financial institutions, given that their primary function is to make loans to creditworthy governments or private sector actors for projects that, at minimum, fulfill the banks' criteria on economic, financial, technical, and legal viability. Yet, MDBs differ from commercial banks in important ways.[8] First, their charters explicitly prohibit them from providing resources that can be made available elsewhere on favorable terms. This is meant to encourage them to complement rather than subsume private sector lending, although in practice it is often difficult to determine whether or not a project would have been undertaken without the MDB's involvement.[9] This is particularly ambiguous when MDBs are able to make private sector loans. Second, MDB loans made to government bodies usually require sovereign guarantees (the EBRD can make nonsovereign loans to public entities). This affects a government's incentives to agree to an MDB loan. Since government budgets can limit the amount of such guarantees that can be authorized each year, these guarantees are likely to be used for priority investments. Third, MDB loans offer longer maturities and grace periods as well as lower interest rates than commercial loans, which enhances their attractiveness. Also, although MDBs finance projects that may be seen as too risky to private sector banks, in some respects they are more risk-averse than private sector banks. This seeming contradiction is explained by the sovereign (and possibly other) guarantees public sector MDB loans take on, which means that what the MDB chooses to finance is backed by the government.[10]

Finally, as development institutions, MDBs are asked to elaborate, incorporate into their lending, and implement an array of mandates given to them by their shareholder member states, to help shape policy reform in recipient countries. A key distinction between MDBs and other development institutions, such as bilateral aid agencies, is that MDBs emphasize loan-making over grant giving, which again shapes MDB activities. Policy changes required by MDB loans can be simple (but not necessarily politically easy), such as requiring municipalities to raise water tariffs, or they can involve more far-reaching reforms in national policy, such as the removal of coal subsidies.

MDBs differ in the degree to which they emphasize their banking goals or their nonbanking goals. Those behaving more like a financial institution than a development agency will be driven more by client demands and contain fewer incentives for staff to situate projects within broader policy goals, such as the environment, than a less banklike MDB. By contrast, a less banklike MDB may face greater challenges in selling investments projects riddled with various policy goals to recipient countries with access to alternative sources of funding with less conditionality attached.

An understanding of how shareholder commitment, organizational dynamics, and recipient demand interact over time requires an analytical framework that examines three stages of policymaking. *Policy objectives* are the MDB's stated environmental goals; *policy process* is defined as the ways these goals are institutionalized into decision-making processes and institutional design; and *policy outcomes* include both the composition of the MDB's portfolio and an assessment of whether its activities are carried out as planned. An institution can essentially become "stuck" at any one of these stages. For example, in cases where donor commitment to the environment is weak—such as at the EIB—we would not expect to see any strong attempts to institutionalize environmental objectives, or much change in the institution's activities. But even a strong policy objective can run into obstacles if it is not well integrated into institutional design and incentive systems. Finally, even in cases where institutions have made consistent attempts to institutionalize new policy objectives, they can still run into trouble at the third stage, if recipient governments or private sector actors do not have incentives to agree to specific projects or the capacity to fulfill their end of the bargain, or if the projects face other obstacles in the implementation stage. A well-designed project may be of little use if it is not carried out properly.

The theoretical argument, which stems from this analytical framework, is that different causal variables matter at different stages of this process. The book will show how neorealist and neoliberal institutionalist approaches emphasizing shareholder politics and preferences work best in explaining how environmental policy goals are brought inside MDBs, and in understanding the depth and scope of an MDB's environmental commitment.[11] However, these theories tend to be static and are poorly equipped to address situations where institutions are given conflicting demands by their shareholders, where institutional governance structures are diffuse, and where there is a mismatch between donor desires and institutional behavior. The book's arguments support the findings of a variety of other institutionalist theories that examine how organizational designs and incentive systems play a crucial role in the translation of environmental objectives into practice, as well as uncovering institutional obstacles in this process. In doing so, these arguments also seek to contribute to versions of constructivist theory that focus on learning processes and the dysfunctional behaviors that can beset IOs, but ignore the ways in which specific institutional mechanisms can help or hinder the translation of new knowledge into actions.

Generally, many approaches to theorizing about the behavior of international organizations tend to separate political and institutional variables without linking the two, and to focus on one of the three stages of the policy process. One of the main arguments of this book is that strengthening our understanding of the linkages between the three stages of the policy process will allow us to more precisely determine the factors influencing an institution's performance. International institutions are being labeled by some scholars as loyal servants of states that continue to remain dominant actors on the international scene, and by others as quasi-independent actors that can shape state interests and sometimes even tell them what to do. The evidence here helps to show that institutional behavior is not an "either-or" issue with respect to the question of autonomy, but rather a "when and how" issue.

The book is structured along the three stages of analysis outlined above—policy objectives, process, and outcomes. Chapter 2 elaborates the conceptual and theoretical arguments and situates them within existing social scientific and policy literatures that offer alternative approaches to explaining MDB environmental behavior. It begins by addressing the issue of how to define the "greenness" of an international financial institution, a topic on which there is neither agreement nor any satisfactory definitions. An examination of the intellectual terrain, in turn, reveals a number of striking gaps, both within and between literatures. While academic debates tend to focus on discrete stages of the policy process, the policy literature offers a range of explanatory variables running the gamut of political and design factors. The policy literature is also overwhelmingly World Bank–centered and hence rarely comparative. It is curious that very little has been written on the environmental behavior of the EIB, because it is hardly a new institution and its lending volume now exceeds that of the World Bank.[12]

Chapter 3 is devoted to policy objectives. It introduces the three MDBs and examines the politics that resulted in their decision to address environmental issues and the resulting depth of their commitment to do so. Tracing the process by which each MDB adopted an environmental mandate, I demonstrate how pressure from major shareholder countries—particularly the United States—usually supported or pushed by NGOs, explains the existence of relatively stronger environmental commitments at the World Bank and EBRD. The absence of these factors explains the EIB's minimal commitment to the environment.

Chapter 4 addresses the policy process, or how the MDBs' environmental objectives were "hard-wired" into their institutional design and incentive systems. It shows how banklike each MDB is through the lens of five indicators: mission, structure, financial tools, major procedures, and "porousness"—openness to outside scrutiny and pressure—and concludes by hypothesizing how variation in the ways the three MDBs juggle their dual personalities of financial institution and development agency may affect their activities in CEE. Chapter 5 turns to policy outcomes. How are environmental issues addressed in the three banks' lending and nonlending activities in Central and Eastern Europe? This account of the three banks' work in CEE provides evidence to support the argument that even where donor commitment to environmental issues is strong, the activities of the MDBs nonetheless reflect how demand-driven they are, while showing how important the differences can be in terms of the MDBs' impact on environmental clean-up and policy reform in CEE. Finally, chapter 6 summarizes the theoretical arguments of the study and draws out their policy implications.

The remainder of this introduction considers some of the fundamental themes and issues of the book in greater depth, and explains its research design and methodology.

MDBs and the Environment

Why does the environmental behavior of an MDB matter? After all, these are institutions whose primary job is to promote economic development and growth, whether by emphasizing poverty alleviation or infrastructural development. There are four main reasons an MDB's ability to incorporate environmental goals into its work is important.

First, MDBs have a substantial impact on recipient countries, through their policy dialogues with recipient governments, their lending in key economic sectors, and their role as a catalyst for encouraging private sector investment. The degree to which MDBs can address local, regional, and global environmental problems through their work influences how these problems are tackled. This has been particularly relevant in CEE, where the fall of the Iron Curtain revealed enormous environmental degradation. At the time, the magazine articles, newsletters, and weekly reports on the region drew attention to issues such as the death of vast stretches of rivers, whose waters were unfit for human drinking or even industrial use; irreversibly damaged forests; and the high degree of birth defects and retarded development in a few seriously polluted areas.[13] As more information was collected and disseminated, it became apparent that much of the environmental damage in the region was concentrated in specific "hot spot" areas, like the "Black Triangle" of southwestern Poland, northern Bohemia in the Czech Republic, and southeastern Germany. It was also clear that enormous investments would be needed to address massive air pollution and major areas of water pollution throughout CEE.

Second, given the many trade-offs between economic growth and environmental protection, the ways in which MDBs attempt to connect diverse mandates offer insights into how ideas of sustainable development can be turned into practice. Happily for the MDBs, there are some areas where the promotion of economic development and environmental protection overlap; these include projects seeking to modernize or build more efficient energy facilities, sectoral restructuring and structural adjustment work that calls for a reduction in energy subsidies, and projects creating financially viable water utilities—wastewater and water supply projects—that reduce levels of water pollution and improve drinking water. MDBs have traditionally been involved in the energy and water sectors, which can make it easy for them to pour old wine into new bottles and recategorize such projects as "environmental" without really changing the way they do business.

However, it is also easy to find examples where the promotion of economic growth and development and the promotion of environmental protection and management clash. These are the projects publicized by environmental groups in their campaigns against the MDBs and include agricultural projects that result in massive deforestation, dam projects for water irrigation that ignore environmental components while poorly addressing resettlement issues, and proposed projects for completing construction of potentially dangerous, partially built (Soviet) nuclear plants.[14]

There is no way to avoid the fact that the promotion of economic development, the principal objective of the MDBs, puts pressure on the environment, since it involves some amount of land clearing, river damming, and increasing use of energy resources. Indeed, the World Bank has admitted that "significant environmental implications" are the rule and not the exception in most of its sectors of lending: in agriculture, transportation, energy, industry, and urban development. The only sectors where projects have relatively minor environmental effects are areas such as education and telecommunications.[15] Even structural adjustment lending (SAL), a tool used by the World Bank that differs from project lending in its linkage to macroeconomic policy reform, is a double-edged sword. Structural adjustment lending may contain conditionality that liberalizes energy, water, or raw materials subsidies, which reduces pollution by lowering demand for the commodity. Yet SAL often results in trade liberalization that can have negative impacts on the environment by increasing commodity exports and thus natural resource depletion. SAL also often requires budget cutting exercises, which can reduce environmental spending.[16]

Third, the dramatic increase of private capital flows to developing countries throughout most of the 1990s forced MDBs to begin rethinking their comparative advantages in a world where private banks are increasingly able to compete with international financial institutions in financing economic development.[17] One result has been a trend by MDBs to focus more closely on linking lending to policy reform, while devising innovative financial mechanisms to undertake their work. The ability of MDBs to cope with environmental issues sheds light on the extent to which they have been able to adapt to new issues more generally amidst a changing global context for their work. The issue of the environment is particularly important as well, because it is the only new policy issue that all MDBs have been seeking to address in recent years, and it has received the most publicity and the most scrutiny from outside actors evaluating MDB performance.[18]

Finally, and perhaps most important, as noted above, there is widespread concern that MDBs have done a poor job in addressing environmental and other new tasks they have taken on, and as a result, must be reformed or restructured. These debates were reinvigorated in 1996, with the fiftieth anniversary of the Bretton Woods conference that gave birth to the World Bank and IMF.[19] Demonstrations against the World Bank and IMF have become larger and more publicized since 2000 as the institutions attract unrelenting criticism from a broader range of antiglobalization groups, which argue that IFI lending has supported undemocratic regimes, failed to improve poverty, and contributed to excessively high levels of debt in developing countries.

Despite efforts to develop new environmental policies, projects, and procedures, as noted above, the MDBs have been widely accused of funding environmentally destructive projects and inadequately addressing environmental issues in their global work. In CEE, for example, environmentalists have attacked the EBRD for lacking an "overarching environmental policy or criteria to guide its project lending," and for being "in some respects even less progressive in its environmental policies and procedures than other international financial institutions."[20] They have also criticized the EBRD for being willing to lend money to complete units of partially constructed Soviet-built nuclear plants in the region, arguing that it is impossible to guarantee the safety of largely Soviet-built models with basic design flaws, even if such loans would be used to bring in sophisticated Western technology. As chapter 5 will show, by the late 1990s, NGOs began to scrutinize the activities of the EIB more closely and were highly critical of its lack of transparency and flimsy environmental portfolio.

Yet, at the global level, MDBs are being called on to play a more visible role in environmental governance. The Brundtland Commission Report, *Our Common Future,* which popularized the concept of "sustainable development" in 1987, argued that multilateral financial institutions "have a crucial role to play" in encouraging environmental investments as an investment in "our future."[21] One conclusion of the June 1992 U.N. Conference on the Environment and Development (UNCED, also known as the Earth Summit, held in Rio de Janeiro) was that billions of dollars of increased funding were necessary to help developing countries pursue environmentally sustainable economic growth, and the MDBs should be an important source of this funding.[22] The World Bank, in particular, has been given a special place in global environmental governance. Three years before Rio, it was asked to design and become the main implementing agency for the new Global Environmental Facility (GEF), to help poor countries address global environmental problems. At the Earth Summit, the GEF was selected to be the interim funding mechanism for the Framework Convention on Climate Change and the Convention on Biological Diversity, the two conventions signed there. The World Bank also manages the Montreal Protocol's Multilateral Fund, which finances the incremental costs developing countries face in phasing out ozone-depleting substances under the Montreal Protocol. The World Bank has been praised by political scientists Peter Haas and Ernst Haas as one of two IOs that have "fully learned to integrate environmental considerations into their traditional responsibilities."[23] Yet, during this same period of time the World Bank has faced harsh criticism from environmentalists who argued that it promotes extensive environmental degradation in developing countries, that it does a poor job in following its own policies, and that it is "intellectually corrupt and institutionally debased."[24] Even its explicitly "green" activities have been criticized. Friends of the Earth, for example, called the World Bank's role in the GEF "a blatant attempt to buy environmental respectability . . . while the Bank continues to (fund) . . . environmentally and socially destructive projects."[25]

Research Design

This book examines the ways in which three MDBs—the World Bank, EBRD, and EIB—have operationalized their relatively new environmental mandates and implemented them in the same geographical region.[26] I analyze the banks' activities in Central and Eastern Europe over the same time period (1990–1999), roughly the first decade of economic transition in the region, during which the banks developed and refined their strategies. Environmental mandate is defined as the authority given to the banks by their shareholding countries to address or emphasize environmental issues in project design or goals, in order to result in greater environmental accountability.

These banks have been chosen as cases for three reasons. First, they have a similar set of dominant shareholders, as table 1.1 shows. The European Union countries are the majority bloc of shareholders in the EBRD, the only shareholders of the EIB, and make up a healthy 26.3 percent of the World Bank's voting power. In fact, Germany, the United Kingdom, France, and Italy, which have the greatest share of voting power vis-à-vis other EU countries in the EBRD and EIB, are the third to sixth largest shareholders in the World Bank, behind the United States and Japan. The United States, meanwhile, is the single largest shareholder in both the World Bank and the EBRD.

Second, the MDBs have broadly similar environmental mandates. On the project level, all three banks are supposed to fund projects with explicit "environmental" goals, such as cleaning up a major river, or sharply reducing emissions of a greenhouse gas. On the policy level, the banks have developed environmental due diligence procedures to use when appraising all projects, as a means of avoiding or mitigating potentially adverse environmental impacts. Finally, as noted above, the three MDBs are major donors in the CEE.[27] While net private capital flows to the region now exceed net flows from MDBs and other donors, the MDBs remain significant actors since they are involved in helping countries liberalize their economies as a means to attract the private flows, while they also leverage investor risk through cofinancing.[28]

I focus on the banks' activities in CEE countries for three important reasons. First, following the collapse of communism in the region, the banks had to devise strategies for economic and environmental assistance to this region. Holding the geographical region and time period constant means that case selection partly controls for these factors. Second, the countries in this region have had to tackle serious environmental problems, of which some of the worst (in terms of health threat) have been airborne dust and sulfur dioxide, and lead found in air and soil.[29] The largest sources of particulate air pollution are low stack emissions from domestic heating and small- and medium-sized enterprises, which rely on low quality lignite, characterized by high sulfur and ash content. Water

Table 1.1
Voting power of G-7 and EU countries within the banks

G-7: % votes	WB[1]	EBRD[2]	EIB[3]
US	16.5	10.1	Capital subscription is not directly linked to voting power, as with the other two banks; for the EIB there are 25 directors, each of which has 1 vote. Germany, UK, Italy, France each have 3 directors. Spain has 2; all other EU states have one; the EU Commission also has one.
Japan	7.9	8.6	
Canada	2.8	3.4	
UK	4.3	8.6	
Germany	4.5	8.6	
Italy	2.8	8.6	
France	4.3	8.6	
Austria	0.7	2.3	
Belgium	1.8	2.3	
Denmark	0.7	1.2	
Finland	0.6	1.2	
Greece	0.1	0.7	
Ireland	0.3	0.3	
Luxembourg	0.1	0.2	
Netherlands	2.2	2.5	
Portugal	0.4	0.4	
Spain	1.5	3.4	
Sweden	1.0	2.3	

European Commission: 3.0
EIB: 3.0

World Bank:		EBRD:	
G7:	42.9%	G-7: 56.6%	
EU component of G-7:	16.7		
EU:	26.3	EU: 51.4	
G-7 + non-G-7 EU:	52.6	EU/EU institutions:	57.4
EU + US:	41.9	EU/EU inst./US:	67.4

1. Data for IBRD only. World Bank, *Annual Report* (Washington, D.C.: World Bank, 1999).
2. EBRD, *Annual Report* (London: EBRD, 1999). Actual voting power is sometimes less than shares. For example, there have been times when the U.S. could not exercise all of its voting rights because its payments to the EBRD were in arrears.
3. EIB, *Annual Report* (Luxembourg: EIB, 1999).

pollution is another serious problem, reflecting inadequate drinking water processing systems and sewage treatment, and unmonitored municipal and industrial discharges, among other factors. Third, CEE is an area where both economic restructuring and environmental clean-up and reform needs were acute during most of the 1990s, making it a region that best challenged an MDB's ability to harmonize economic and environmental strategies.

MDB lending to CEE countries addressed in this book began earlier than lending to other countries in the region, such as the largest former Soviet republics, and as a result, there is a greater body of MDB activities to compare in the former group. However, it must be noted that the recipient countries all three MDBs have in common constitute a smaller set of countries. While the EBRD and World Bank also lend to the former Soviet Union, as of 2000, the EIB's lending to former Soviet countries was limited to the Baltic states. The EIB also first extended lending to Bosnia and Herzegovina and Macedonia/FYR in 2000 but had not yet initiated lending to Croatia.

The book compares the three banks at the three stages of policymaking, examining how different explanations might account for the significant variation in MDB environmental performance in CEE. In effect, the book traces the processes by which new environmental mandates appear in the banks, are institutionalized, and are translated into specific sets of activities in CEE. Analyzing the observable implications of different theories and arguments seeking to explain MDB behavior at each stage of the policy process increases the number of observation points for investigation. While an in-depth analysis of the implementation of all 500-plus MDB projects in CEE is beyond the scope of the book, I examine available evidence that underscores some of the difficulties the banks have had translating project ideas into practice and measuring a project's environmental impact. This evidence includes written evaluations and primary and secondary materials on the banks' projects. I also interviewed over 100 MDB officials, environmentalists, project managers, ministry and municipal officials, and consultants in Washington, D.C., London, Luxembourg, Poland, Hungary, the Czech Republic, and the Baltic states. The findings suggest that despite numerous sources of institutional inertia and the complexity of defining an MDB's environmental behavior, there is a clear range of activities these institutions are able to undertake that at least avoid significant environmental harm and at most may directly contribute to environmental improvement of priority issues. A better understanding of the factors that help or hinder their ability to undertake these activities is vitally important at a time when all MDBs continue to take on new issues each year while calls for their closure become more persistent.

2

Intellectual Context: Understanding MDB Greenness

How does one define the "greenness" of international financial institutions such as multilateral development banks? What factors shape their environmental behavior? In the midst of the debates about these institutions, answers to these questions are surprisingly unclear. This chapter addresses these questions. It begins with a conceptual discussion of the challenges posed in attempts to define or measure the environmental behavior of MDBs. The chapter presents a simple continuum as a way of delineating MDB environmental goals and argues for the importance of analyzing an MDB's environmental behavior at different stages of the policy process. It then situates these arguments within the broader literature seeking to explain the behavior of international institutions in general, and the environmental behavior of MDBs in particular, highlighting some shortcomings of existing explanations.

The book's central argument is that analysis of the process by which MDBs receive, institutionalize, and implement environmental policy goals highlights the sources of inertia, as well as innovation, in their behavior. Shareholder pressure, usually pushed by NGOs, is a key factor determining the degree to which an MDB seeks to address environmental issues in its work. Stronger environmental commitment has not been internally initiated in any of the three MDBs. In cases where strong shareholder pressure is absent, such as the EIB, we would not expect too much on the performance side. Yet, even where shareholder pressure is significant, MDBs struggle with their environmental performance. This is owing to a sometimes poor or awkward fit between new environmental mandates and other institutional design characteristics and incentive systems. As a result, neither shareholder preferences nor institutional design and incentive systems alone are sufficient to explain MDB environmental behavior. Analyzing the factors that shape the ways new environmental issues are brought into an MDB and are "hard-wired" into its institutional processes and activities provide a causal sequence that offers a more precise explanation of the process by which institutions adopt and implement new mandates. From a policy perspective, a better understanding of these factors can provide points of leverage for those seeking to influence MDB behavior.

This process-based approach exposes important ambiguities in the existing literature. International relations and comparativist scholars studying institutional behavior, for example, tend to focus on one stage of the policy process, in part because they are asking different kinds of questions. Neorealists and neoliberals are more interested in why international organizations exist, how they provide a forum for states to cooperate, and what impact institutions have on state behavior than in analyzing institutional performance and the mechanisms that shape institutional activities and outcomes. Some comparativist and constructivist scholars, in turn, focus more on issues such as institutional autonomy and the role of institutions in providing information and shaping norms. At the same time, the specific literature on MDB environmental behavior offers a variety of explanations that, while compelling, are rarely backed by rigorous analysis or methodology. This is not surprising, for example, in the work of environmental advocates seeking to change MDB policy or activities by publicizing an MDB's problem projects, versus a broader perspective that includes analysis of successful projects or places problem projects in the context of an MDB's entire portfolio. There are also gaps between the bank-centric policy literature, and the broader scholarly literature, with each addressing explanations not well examined in the other.

Defining Environmental Behavior

Defining the environmental behavior of any institution is a difficult conceptual exercise because the boundaries of what is and is not an "environmental" outcome are not always clear. The entanglement of the concept of "environment" with that of "sustainable development" or "environmentally sustainable development" only increases the definitional opaqueness of "environmental" behavior. As Ismail Serageldin noted, from the position of World Bank Vice President for Environmentally Sustainable Development, "What is sustainable development other than sound economic management, rationality, respect for the rights of others, and concern for future generations?"[1] Environmental behavior can also be found in a range of activities, such as project financing, bank policies, research products, partnerships or networking arrangements, and agenda-setting exercises such as conferences that can be difficult to measure in terms of environmental impact, especially when they involve indirectly shaping political processes, dynamics, or norms.[2]

The most common approach the banks themselves use to define their environmental behavior is to emphasize specific activities with clearly defined primary environmental goals, such as pollution abatement or cleanup, or conservation. The focus of such projects is the provision of environmental benefits. However, discrete environmental financing presents only a part of the picture of the banks' environmental behavior, even if the focus is on their lending activities. It is important to know whether they are simultaneously undertaking a number of potentially environmentally harmful projects. Likewise, a bank may have few explicitly "green" projects, but the bulk of its portfolio may not be environmentally damaging; for example, it may consist of educational or telecommunications projects. It may not be actively pursuing environmental goals, but neither is it doing any environmental harm. In fact, some projects that are not designed with specific environmental components may have a positive impact on the environment, such as projects to modernize industries that result in more efficient use of energy and less pollution. A bank may also have few environmental projects, but it may make many adjustments in its wider portfolio of projects to avoid or alleviate environmental damage. It is also important to know if the bank's environmental projects bear any relationship to the region or country's most pressing environmental problems.

Moreover, to what extent is the size of a project important? There is often no way to know precisely how the loan amount correlates with the environmental benefits of a project. There is anecdotal evidence that some of the smaller projects the MDBs fund are the most effective in terms of their environmental benefit, in line with evidence that more complex projects are often more difficult to implement.[3] It is also sometimes difficult to quantify how much of a project loan is being used for environmentally related components, and what the actual environmental *impact* of a project is.[4] In addition, an environmental component of a project may be small in monetary terms, but large in terms of its impact on the environment. Equally important, data on the environmental components in a project's design does not tell us anything about how a project has been implemented, and implementation problems are indeed a key focus of NGO campaigns and criticism.

Another difficulty in categorizing MDB projects in terms of their environmental goals is that most MDB projects have multiple objectives, which means there may be no precise way to assess the weight of one objective versus another. Where objectives are quantified in terms of the share of environmental costs or benefits, this methodology is not a precise science and is based on numerous assumptions that may or may not be correct.

The nature of the environmental components or goals of a project depends, in part, on the sector. In the transport sector, for example, environmental components or goals may involve the development of mass transit, traffic management strategies to reduce congestion, or policies that encourage higher occupancy vehicles or increased vehicle fuel efficiency. Environmental components or goals of water projects may include water quality management, pollution abatement, and the treatment and disposal of wastewater and sewage. Water sanitation projects tend to be automatically considered "environmental" projects by all three banks, but whether or not a water treatment plant relies on mechanical treatment versus biological or chemical treatment can greatly influence its environmental impact. Water supply projects are more difficult to categorize, since they can be environmentally harmful if they involve diverting rivers or draining aquifers in an unsustainable manner.[5] Indeed, water projects often require partial environmental assessments (EAs) at the World Bank and EBRD, since these projects have important environmental impacts.

In the energy sector, an MDB's environmental mandate is reflected in the degree to which its projects address supply-side efficiency gains, demand-side management and conservation,[6] the promotion of renewable or relatively cleaner energy sources, as well as the reduction of emissions through end-of-pipe technology (such as flue gas desulfurization—or FGD—at power plants). In forestry, a "green" project would clearly include forest protection and conservation management components.

The MDBs themselves do not share similar definitions of what their "environmental" work is, in terms of the projects they finance. Even within one MDB there may be multiple definitions. These definitions have also evolved over time. They are further discussed in chapter 3, since they are interwoven with the development of MDB environmental policy. Generally, the EIB and EBRD define their environmental project work in an ex post facto manner to describe the types of infrastructure projects they normally undertake. The World Bank has made a more explicit effort to define what it means by environment, and to produce research that examines the relationship between growth, poverty reduction, and environment.

The EIB's preferred method of defining its environmental behavior, for example, is to list types of projects, namely, traditional MDB projects that happen to have some positive impact on reducing emissions or using resources more efficiently. These include projects in the water supply and wastewater treatment sector, solid waste treatment, projects that reduce air pollution (especially from power stations), and those that involve "the protection of the environment and the improvement of the quality of life in urban areas."[7] Curiously, its most recent environmental policy document does not mention some types of ostensibly environmental projects the Bank has funded in the past, such as reforestation work undertaken in Italy, Spain, the United Kingdom, and Portugal.

The EBRD also does not explicitly define its environmental lending behavior, and like the EIB, it tends to provide examples of the types of environmentally oriented operations it funds, which are primarily through its municipal infrastructure and energy efficiency groups and include water supply systems, wastewater management, solid and hazardous waste management, as well as district heating, energy efficiency projects, renewable energy, and urban transport. Since the early 1990s, the World Bank has sought to define more clearly the environmental nature of individual projects. A 1994 document defines environmental projects as "those that are undertaken largely for purposes of environmental protection, conservation, rehabilitation, planning, management, education, or institutional strengthening."[8] It also distinguishes projects with "primary environmental objectives" from projects with "major environmental components" or simply "environmental components." The former are projects where the cost of environmental protection or

environmental benefits exceed fifty percent of total project costs or benefits, and tend to include projects in forestry, water supply and sewerage,
and environmental capacity building, as well as some energy rehabilitation projects. Projects with major environmental components are those
where the environmental costs or benefits are less than fifty percent. The
World Bank also distinguishes between "brown" projects, or projects
that emphasize clean-up or abatement, "green" projects, or those that
focus on natural resource management and institutional development
projects. Given its character as a lending institution, it is no surprise the
Bank emphasizes "brown" projects over "green." "Targeted programs
for the environment," according to the Bank, usually consist mostly of
pollution management and urban environment projects, followed by rural environmental and institution-building projects.[9]

Yet even the World Bank's technique is more indicative than precise.
The calculation of environmental costs and benefits is not a science. In
addition, some of the so-called environmental projects have had dubious
environmental merits, including a variety of forestry projects criticized
for their promotion of logging and emphasis on timber exports.[10] Sometimes the Bank itself has no unified view on what an "environmental"
project is, evidenced by the different definitions and different categories
appearing in various Bank documents and publications. In the case of
CEE, there have been a number of versions regarding total World Bank
environmental assistance to the region. Just focusing on the year 1994,
the Bank's annual report that year listed only one loan ($18 million) in the
environmental sector for the region for the period 1990–1994.[11] Another
publication it produced added two other projects to the category of projects with "primary environmental objectives," totaling $184 million.[12]
Still other World Bank documents include certain structural adjustment
loans that contain environmental components, which pushed the total up
by another $1 billion.[13] Another document listed a total of six stand-alone
environmental projects for the period 1990–1994, totaling $1 billion.[14]
Projects with environmental components or objectives in the same period
add an additional $517 million to the total and do not include the envi-
ronmental components of SALs. Jumping ahead to fiscal 2000, the World
Bank lists its total for environmental lending to the region as a paltry
$384 million.[15]

Table 2.1
Continuum: Range of MDB environmental objectives

Minimal Activity "Light Green"	Environmental Additionality	Deeper Activity "Dark Green"
• Project design seeks to mitigate environmental harm	• Projects have significant environmental objectives or benefits	• Projects have primary environmental goals; other technical assistance or agenda-setting activities in environmental development

This discussion illustrates the challenge of defining "environmental behavior" in terms of discrete projects financed, but other issues arise when broadening the definition of behavior to include the depth and scope of the bank's environmental policies and additional nonproject activities it may undertake, including agenda-setting activities or research. These issues are explored further in subsequent chapters.

Table 2.1 proposes a simple continuum to categorize MDB environmental *objectives* (versus behavior or performance in individual projects or activities). On the left end of the continuum is a minimal degree of commitment or activity, where the MDB seeks to guarantee that its projects are not environmentally harmful, in the sense that it seeks to avoid or alleviate possible degradation of a country's natural resource base, or biodiversity loss. Obviously, exactly where the line is crossed between environmentally harmful and acceptable is often a gray area, viewed differently among different actors. The point is that the MDBs may continue financing traditional types of MDB projects, but the banks show signs of positive environmental behavior if they make sure these projects are properly appraised to ensure that environmental problems are averted. This mitigation exercise is usually carried out through the MDB's environmental assessment procedures and its rules about what types of environmental standards to follow in its work.

Moving toward the right end of the continuum, policy objectives start to include actively seeking projects with significant environmental objectives or components. In starting to be more active in explicitly addressing environmental issues, the MDB may also more broadly consider natural resource management issues in its work and engage in measures that shape environmental policy development. This involves explicit attempts by the MDB to influence the recipient government's environmental policy, whether through capacity-building exercises, persuasion, or explicit conditionality.[16] On the right end of the spectrum we see the "darkest green" behavior, where the MDB finances projects with primary environmental goals and attempts to integrate environmental thinking into the broader set of strategic goals it develops for each recipient. In other words, rightward movement reflects attempts by the institutions to innovate in ways that transcend their traditional actions. The movement along this continuum is additive. As subsequent chapters will show, the EIB clearly is situated toward the "minimal," "light green" end of the spectrum, the World Bank toward the "dark green" end, and the EBRD somewhere in the middle. Yet, if we take a longer, historical view, we would see that the World Bank began its life on the left end of the spectrum and has moved rightward over time. The EBRD has also moved rightward over time, but not to the same degree as the World Bank. The more complex factors that explain shifts in the banks' positions, as well as how commitment translates into specific lending and activity decisions and performance, are topics of subsequent chapters.

Explaining MDB Environmental Behavior

A process-oriented approach posits that institutional behavior is best understood by tracing new institutional mandates through different stages of the policy process to better determine when and how external political and internal institutional factors shape policy outcomes. The policy process itself—from the identification of new mandates to their implementation—offers a set of causal mechanisms linking new ideas to on-the-ground practices. A comparison of the three MDBs at different stages of the policy process highlights sources of innovation and inertia in their environmental behavior by revealing where behavior is "sticky." Such an analysis contributes to institutionalist approaches that seek to explain IO behavior, while challenging approaches that focus on the preferences of major state (donor) actors, and those that address that issue of how ideas enter institutions and are embodied into procedures versus what happens to these ideas as they are translated into institutional performance.[17] The empirical evidence for these arguments is found in the following three chapters. The remainder of this chapter examines and evaluates alternative theories, highlighting some of the gaps in their coverage.

In general, writers specifically analyzing MDBs and the environment are highly critical of the banks' ability to effectively tackle environmental issues in their policy and project work. Most of these articles lay out an argument or series of arguments that are backed by selective illustrations. While generally methodologically weak, this literature does address the issues at the heart of this book and it proposes explanations that find parallels in the broader institutional literature.

This work is generally produced by policy analysts and environmentalists, as well as a few political scientists. It is overwhelmingly preoccupied with the World Bank, on which where there are over a dozen recent articles and a few books analyzing the Bank's environmental behavior.[18] By contrast, there are a handful of articles and reports analyzing the EBRD, and even fewer on the EIB and its environmental behavior, and to date, little comparative work on MDBs.[19]

It is not surprising that so much more critical attention is focused on the World Bank. The Bank is the oldest MDB, with the broadest (global) membership and scope of activities. It is a leading actor in global thinking and research about development issues. There is a vast literature on the Bank and its role, activities, and ideas in promoting development in the Third World. From an environmental perspective, the Bank also has had a series of spectacularly disastrous projects, which sparked a major campaign by NGOs in the 1980s to force the Bank to reform its environmental practices.[20]

The fiftieth anniversary of the 1944 founding of the Bretton Woods institutions—the World Bank and the IMF—inspired a flurry of assessments on the roles of the two institutions and the steps that should be taken to help them better adapt to the challenges of the twenty-first century. The G-7 major industrialized countries also called for a reassessment of MDBs, in ways that encourage them to invest more in people, public participation, and the environment.[21] A group of environmental, development, human-rights, and other NGOs, in turn, launched a "Fifty Years is Enough" campaign, calling for far-reaching changes in both institutions in the areas of public accountability, the emphasis of lending programs, and the way environmental problems are addressed. The Asian financial crisis, sparked by events in Thailand in 1997, intensified debate on how to reform the two Bretton Woods organizations as well as the broader global financial architecture. NGO criticism of both institutions escalated at the IMF/World Bank April 2000 spring meetings, when thousands of protesters descended on Washington, D.C. decrying the institutions' actions on a wide array of issues, including social justice, debt, and the environment. Indeed, the Bretton Woods institutions, along with the WTO, became the focus of a wide variety of criticism against processes of globalization.

One could hardly expect the same amount of debate to surround the EBRD, which was established only in 1990, and which, with its focus on former communist countries, avoids many of the prickly north-south issues that touch the World Bank. Yet, as noted above, it is curious to discover the dearth of literature on the EIB and its environmental behavior, given the bank's age (created in 1957) and its large lending volume. Although 90 percent of the EIB's lending is in Western Europe, it is still active on a global scale, and even its projects within Europe obviously have an impact on the environment.

The literature focusing on the World Bank is highly critical. Only a few brave souls (outside the Bank) offer praise, and do so by emphasizing changes in Bank policy and procedures, but without an examination of how these policies are (or are not) translated into outcomes in terms of project design and implementation. Political scientists Haas and Haas, for example, in analyzing the response of thirteen IOs to the "environmental problematique," found the World Bank and UNEP to be the only ones that exhibited traits of "learning" in the ways in which they integrate environmental issues into their work.[22] Learning is defined as a cognitive evolution, where consensual knowledge is used to define and solve problems. They argue that it is rare for learning to take place in IOs; instead, most organizations merely "adapt," by changing procedures or routines without a deeper examination of underlying values.

As evidence of the World Bank's learning, they describe the growth in its environmental procedures, activities, and staff that has taken place since 1989. In particular, they stress the importance of the growth in the number of environmental specialists within the Bank, who have been placed inside the operational (lending) divisions. Epistemic communities seem to be a driving force in prompting institutional learning, and the authors list various institutional characteristics that explain why these communities are better able to penetrate some IOs rather than others. Characteristics of "learning institutions" include an open flow of information from scientific communities or NGOs, relatively autonomous secretariats, and influential shareholders from countries possessing democratic cultures.

Although Haas and Haas make a contribution in identifying pathways of policy evolution in an institution, absent from their argument is an examination of the impact the learning has had on what the World Bank *does* in the field, in terms of how it implements its projects and procedures. Indeed, critics of the Bank's environmental behavior tend to discount the evolution of policies and procedures, arguing that they have made little difference in terms of the Bank's on-the-ground performance. These writers criticize the Bank for funding projects that caused enormous environmental degradation, for having weak environmental policies, and for failing to follow the policies it has developed. The World Bank itself is aware of its poor record in implementation and has sought to strengthen its monitoring mechanisms.

One of the harshest critics of the World Bank's environmental behavior is Bruce Rich, a senior attorney at the Washington, D.C.-based NGO, Environmental Defense. In his 1994 book, *Mortgaging the Earth: The World Bank, Environmental Impoverishment, and the Crisis of Development,* Rich highlighted case after case of World Bank projects he argued were environmental debacles, which destroyed coastal ecosystems, promoted (directly or indirectly) enormous deforestation, exhausted soils, and displaced thousands of people.[23] Among the more publicized failures was the huge Polonoroeste project in Brazil, in which the Bank provided over $450 million in loans in the early 1980s to promote agricultural colonization and road-building in the state of Rondônia. According to Rich and others, the project encouraged a massive migration of colonists that overwhelmed support efforts, resulted in slash and burn agriculture, and ultimately was responsible for enormous deforestation, among myriad other problems.[24]

Despite the World Bank's adoption of a wide range of new environmental policies, guidelines, and procedures beginning in the late 1980s, it continued to fund environmentally destructive projects, and critics have pointed out numerous cases of where, they argued, the Bank was not following its own internal policies. One of the most contentious projects of the 1990s, for example, was the Sardar Sarovar dam project in India, which provoked enormous outcry from environmentalists.[25] After criticism of the projects spread to include the opinions of U.S. congressmen and senators, as well as legislators from Japan, Finland, and Sweden, World Bank president Barber Conable asked former U.S. congressman and former director of UNDP Bradford Morse to undertake an independent study. The resulting 1992 report confirmed earlier criticisms.[26] It found that the projects were poorly appraised, and that the Bank was not enforcing its own policies on resettlement or environment. The involuntary resettlement resulting from the projects, said the report, "offends recognized norms of human rights."[27]

With respect to the projects treatment of the environment, the report was highly critical. "The history of the environmental aspects of Sardar Sarovar is a history of non-compliance," it said.[28] "There appears to have been an institutional numbness at the Bank and in India to environmental matters," it continued. "The tendency seems to have been to justify rather than to analyze; to react rather than anticipate."[29]

This project was not alone in generating a great deal of negative publicity and criticism of the World Bank's environmental behavior. Another controversial project was the proposed Arun III hydroelectric dam project in Nepal, which was ultimately canceled by World Bank president James Wolfensohn in 1994.[30] The Bank's forestry projects and energy projects have also been attacked for failing to follow the Bank's own policies.[31]

Going beyond purely environmental debates, but certainly affecting them, was the 1992 in-house report commissioned by former World Bank president Lewis Preston to review the quality of the Bank's portfolio. The Bank's review team, lead by Willi Wapenhans, Preston's special advisor and a Bank vice president, found that the Bank's portfolio had steadily deteriorated over the years. Projects with "major problems" increased from 11 percent to 20 percent between fiscal 1981 to fiscal 1991. Citing the Bank's Operations and Evaluations Department, the report said the number of "unsatisfactory" completed projects increased from 15 percent of the sample reviewed in fiscal 1981 to 37.5 percent in fiscal 1991.[32] Project completion on average was almost two years over predictions. Borrowers' compliance with the legal convenants written into the loan agreements was "startlingly low." "Loan agreements," said the report, "do not induce the behavior expected and their credibility as binding documents has suffered."[33] An extensive in-house review of 14 sectors by the Bank's own Quality Assurance Group in 1997 blamed some portfolio problems on borrowers, but saved its most scathing criticism for the Bank. While borrower factors included weak or no government commitment, inadequate participation by beneficiaries and broader macroeconomic instability, Bank-side problems included issues such as staff overoptimism and overambitious projects. "Institutional amnesia is the corollary of institutional optimism," it said.[34]

Other critics focus not on individual Bank projects, but rather on the Bank's emphasis on free-market policies, particularly its structural adjustment lending (SAL), which calls for the promotion of conservative fiscal and monetary policies to help developing countries correct often deeply rooted domestic economic distortions. Critics argue that these loans, by liberalizing trade, promote exploitation of natural resources for export. Furthermore, the Bank's SALs as well as the Bank's programmatic lending are not subject to the Bank's environmental-related policies, such as environmental impact assessment.[35]

To summarize, the bulk of criticism of the World Bank's environmental behavior has tended to focus on a set of disastrous or potentially disastrous projects, other examples where the Bank does not follow its own policies, the negative side-effects of SALs, and evidence of poor project implementation, generally. Since the focus is on publicizing problem projects or highlighting weak policies, it is unsurprising that there is no analytical work outside the Bank that looks at a larger sample of projects, to determine what percentage of the overall portfolio is environmentally problematic. Are the problem projects the exception or the rule? It is also not clear that SALs are unilaterally negative with respect to the environment, since research on this topic has produced mixed results.[36] Nor is there much outside work seeking to determine the extent to which change has taken place in the Bank's environmental behavior since it adopted a series of new environmental policies beginning in the late 1980s.[37] One important exception is a detailed study by Robert Wade, who argued that the Bank's attempts to translate new environmental procedures into practice between 1970–1995 were hampered by the "slipping clutch" of the Bank's incentive system. Among the numerous gaps in incentives he highlights are measures of individual performance that are more procedural than outcome-oriented and budgetary and rewards systems that hinder a successful "mainstreaming" of environmental management into the Bank's system. [38] More recently, the Bank's own Operations Evaluation Department produced a report in 2000 that examined the Bank's environmental performance since the late 1980s. It concluded that "achievements . . . have lagged behind expectations" given the existence of "gaps, inconsistencies and ambiguities in the Bank's approach toward promoting environmental sustainability."[39] It pointed out numerous weaknesses in Bank policy, including the lack of a comprehensive environmental strategy, an environmental assessment framework that does not take into account the Bank's programmatic or structural lending, which is around half of the Bank's total lending, and gaps in supervision of project implementation.[40]

The environmental behavior of the EBRD and EIB has also been criticized by environmental groups, but on a far smaller scale. In general, both banks have been attacked for doing a poor job of adhering to environmental procedures that the NGOs argue are already weak. In its 1992 report on the EIB's environmental behavior, in one of the few published reports to date on that topic, WWF International argued that the Bank's environmental procedures are informal and subjective, with no well-defined framework for outlining conditions for lending.[41] More recently, CEE Bankwatch Network, a network of NGOs throughout the region, argued that the EIB's environmental performance is weaker than that of the World Bank and EBRD, in part because the EIB lags behind its sister banks in terms of transparency and accountability.[42]

The Center for International Environmental Law, in turn, criticized the EBRD for lacking an "overarching policy or criteria to guide its project lending," and for not adequately following its own procedures in a number of its projects. One controversial example behind this criticism was the case of the 1995 proposed EBRD loan to complete units 1 and 2 of the Mochovce nuclear plant in Slovakia, which was 90 percent built by the Soviets. The proposed 412.5 million Deutsche Mark loan would have been the biggest loan in the Bank's four-year history, but it stirred enormous controversy.

Opponents, including environmental groups, the Austrian government, the U.S. Executive Director of the EBRD, and some officials within the European Commission and even the EBRD, argued that it is impossible to guarantee the safety of a largely Soviet-built model with basic design flaws, even if the loan would be used to bring in sophisticated Western technology. Indeed, Austria threatened to withdraw its membership from the EBRD if the loan were approved. Opponents of the loan argued that the Bank's own economic justification of the loan as a "least cost option" was highly ambiguous, and thus the loan should be dropped as not meeting the Bank's own environmental and lending conditions. The Bank's "least cost study," showing that the completion of Mochovce was cheaper than building a combined-cycle gas turbine plant, was criticized as basing its conclusion on unrealistically high forecasts of gas prices and discounting rates for decommissioning.[43] Ultimately, the Slovaks threw a wrench into the debate by deciding to bypass the Bank and seek funding on their own from Russia and a Czech firm to complete the plant for one-third of the cost proposed by the EBRD.

In more recent years, environmentalists have remained critical of what they see as the EBRD's lack of strategic vision on the environment.[44] They are particularly disappointed with the EBRD since it was the first MDB with the goal of promoting sustainable development built into its founding articles. To date, however, neither the EBRD nor the EIB has been accused of being involved in Polonoroeste-type debacles.

The policy literature on the environmental behavior of MDBs offers a number of arguments to explain this behavior. In terms of the World Bank, some writers focus on a single factor, such as corporate culture, or the role of the Bank's president, or the Bank's emphasis on structural adjustment lending. Yet more often, existing explanations blame the Bank's struggle to implement its environmental policies on a host of factors, and it is difficult to untangle one potential cause from another, or to determine the importance of one factor versus another. The bulk of these explanations tend to fall into three categories: governance, incentives, and what Naim calls "goal congestion."[45] Often, these categories overlap, although causality is rarely specified. Many, but not all, of the debates raised in the policy literature are reflected in the broader theoretical debates on institutional behavior, which add depth and offer additional arguments to the former.

Governance

While some earlier work on the World Bank argues that it is essentially the creature of the developed capitalist countries (particularly the U.S.)

to promote their own political and economic interests,[46] much of the more recent work makes the argument that governance is in fact quite weak. The general argument is that the Bank's 24 executive directors represent over 180 countries and have no real control over the Bank's projects and policies. Board weakness is thought to be a result, in part, of structural and design-related variables: that is, there is high turnover, with directors appointed for renewable two-year periods, staying, on average, for three years; and most directors represent a number of countries (up to 25), which dilutes their ability to act on country preferences. The fact that the board brings together countries that are capital providers and those that are capital users also creates a situation where country preferences often differ, which provides disincentives to cooperation on divisive issues. In an example of "bloc loyalty," borrower countries tend not to vote against loans for each other. Some critics argue that the board is weak by choice, that is, as the political outcome of a tacit agreement between countries of the north, who seek to use the Bank as a way to address north-south relations; and countries of the south, who are "addicted" to the Bank's loans.[47]

No matter what the cause of board weakness is, the result is a relatively autonomous bank management and staff that lack incentive to be accountable to the board. As Naim, a former executive director of the World Bank, argued, "A divided board of overwhelmed directors, many of whom cannot afford to irritate the Bank's management, and usually leave by the time they begin to be more effective, is no match for a usually brilliant group of professionals with decades of experience at the Bank."[48]

A more extreme view of this argument is that the Bank has become a bureaucracy gone wild, a sort of "World Bankenstein" lacking accountability to shareholder countries.[49] George and Sabelli, in turn, call the World Bank a "total social phenomenon, a Thing," comparable to the medieval Church with "a doctrine, a rigidly structured hierarchy of preaching and imposing this doctrine and a quasi-religious mode of self-justification."[50] Other executive directors of the World Bank have been quoted as saying (in frustration over the Sardar Sarovar dam project) that the Bank has shown "a profound lack of accountability to its shareholders" and has suppressed "information on controversial projects."[51] Therefore, even if major shareholders push for the Bank to better address environmental issues in its work, according to this argument, Bank management and staff can easily ignore the pressure.

The degree to which MDB management and staff are accountable or not to the bank's shareholders reflects broader academic debates about the sources of behavior within IOs. IOs can be expressed as a classic principal-agent relationship; they are created by states as a means of overcoming collective action problems, promoting cooperation, and creating advantages by delegating to agents the responsibility of carrying out their interests.[52] Yet agency opportunism may exist in any principal-agent relationship and the fact that both the principals and the agents seek occasions to maximize their interests creates an inherent tension between the two. This tension is likely to be intensified in organizations like MDBs, where principals have multiple and often ambiguous preferences, while the institution itself is responsible for distributing billions of dollars of resources.

The debates over who drives institutional behavior—principals or agents—are often presented as debates over how much autonomy agents exhibit when they carry out the activities delegated to them by their principals. The above view that bank management and staff are largely autonomous would be readily accepted by a number of historical institutionalist and constructivist scholars, who argue that IOs, once established, can "take on a life of their own and develop their own internal dynamics," or emphasize unintended consequences or path dependence in analyzing institutional development.[53] In this constitutive perspective, IOs can also help define and even transform state preferences, as well as act as "autonomous sites of authority."[54] Some scholars specifically using principal-agent frameworks to analyze regulatory politics also subscribe to the "runaway bureaucracy thesis," or situations where lax principals give regulatory bureaucracies the opportunity to pursue their own goals.[55] Yet, scholars analyzing institutional autonomy are also aware that shareholding countries have the legal ability to reduce agency slack, or "rein" in the agents, if they choose to. Institutions can structure and shape the outcomes of political forces and battles, but may not be the primary cause of outcomes.[56]

Structural realist and neorealist international relations theorists, on the other hand, would strongly disagree with the weak governance argument above. Agents may have meaningful "slack," but this is a reflection of them doing their job in using their expertise to carry out what they believe the principals want them to do. Common opportunities for agent opportunism, such as hidden information and hidden action, are dealt with through contract design, monitoring mechanisms, and institutional checks.[57] IOs, in this view, are essentially the passive instruments of states, and they have power and agency only to the extent that this is tolerated by their owners.[58] "Institutions," wrote Mearsheimer, "have minimal influence on state behavior, and thus hold little promise for promoting stability in the post Cold–War world."[59]

Realist perspectives have come under sustained attack by several groups of scholars in recent years. Neoliberal institutionalists, for example, argue that international institutions do more than simply reflect powerful state interests; they also provide important functions that transcend the simple Prisoner's Dilemma problem and encourage international cooperation. In Keohane's classic statement of this functionalist view, institutions establish patterns of legal liability, provide more symmetrical information, and reduce transaction costs—the costs of bargaining, side payments, and reaching agreement on common issues. They do so by providing fora where states can meet repeatedly, which facilitates negotiations and thus promotes cooperation.[60] Nonetheless, while neoliberal institutionalists admit that institutions may be more than passive actors, they do not believe institutions have the ability to make authoritative decisions to bind states.

More recently, international relations scholars have begun to examine different organizational structures and forms, as a way of determining how a chosen form helped member states achieve their interests. Less attention is paid, however, to the impact organizational forms themselves ultimately have on policy outcomes, how existing institutions adapt to new policy mandates, and other questions that go beyond the basic question of autonomy.[61] As Martin and Simmons noted in a review article on theories of international institutional behavior, "Too often over the past decade and a half the focal point of debate has been crudely dichotomous: institutions matter, or they do not."[62]

The empirical chapters that follow show that both the weak governance and neorealist arguments fall short in explaining MDB environmental behavior. A state-centered perspective appears to have some explanatory power, since there is a strong correlation between the presence of the pro-environment United States on the bank's board and the degree to which the bank attempts to tackle environmental issues. Such an explanation may also account for why the EIB—where the capital providers *are* the recipients of the vast majority of the Bank's lending—does not face pressure to pursue environmental conditionality in its work. Stated differently, why would its shareholders want their bank to tell them what to do?[63] In addition, there are clearly instances where shareholder countries have reined in the banks, when agency slack became uncomfortable. For example, in response to controversy over the Sardar Sarovar project, major donors like the United States pushed the World Bank to set up an independent inspection panel to increase accountability by investigating complaints from people affected by the Bank's projects. The EBRD board, in turn, fired the Bank's first president, Jacques Attali, owing to disappointment over his performance.

However, these explanations do not account for many important differences between the banks. In general, realist and neorealist debates revolve around the analysis of when and how states cooperate, in this case through multilateral fora. As a result, these scholars have the most to say about factors in the inter-state bargaining and negotiations process that lead to regime or IO formation and changes, but pay less attention to what happens *after* the IO agents have received their delegated responsibility, how their actual policy actions reflect or do not reflect these responsibilities, and what other factors may have an impact on how these are carried out. Realists would not see these aspects of the policy process as important as long as outcomes reflect the general tasks the agents are supposed to do. Yet IO agents such as MDBs often operate with conflicting mandates, while institutional design and incentive systems greatly influence the depth and scope of an MDB's response. There are also examples where the creation of policy activities undertaken by the World Bank that have changed the ways its major donors shaped their own bilateral assistance to CEE.[64]

In addition, chapter 4 will show how the governance structures of the MDBs make it difficult for any state shareholder to translate its power into influence, that is, that weak governance *is* a characteristic of the World Bank, as well as the other two banks, and that it does have an impact on institutional incentives. Thus, it is important to show *when* and *how* shareholders exert influence on the bank's environmental behavior, and when and how they do not.

Other Institutional Incentives

Attention to the institutional mechanisms that shape IO behavior has strands both in the policy and academic literature on MDBs, as well as the broader social scientific literature. In the former, Rich (in addition to his weak board argument) emphasizes lending pressure as the primary factor discouraging World Bank staff from paying close attention to environmental issues. The pressure to make loans is attributed to specific lending targets that were initiated under Robert McNamara's stewardship of the World Bank (1968–1981), demand for loans from developing countries, and the always-present pressure for staff to find financially attractive, "bankable" projects. This became a sensitive issue for the World Bank in the late 1980s when it began receiving more payments from LDCs than it was making in new loans to them.[65] The pressure to get loans out, writes Rich, is far stronger than any pressure for staff to be accountable for how those loans are implemented.[66] Indeed, he argues that loan repayment has no connection to the quality of project implementation, since countries must repay the Bank whether or not the project is well implemented.

The Wapenhans Report, in explaining the World Bank's poor performance, also pointed to "the Bank's pervasive preoccupation with new lending," which it noted some have dubbed an "approval culture."[67] Quantity, it argued, has become more important than project quality in the Bank's history.

Other analysts turn their attention to the issue raised earlier in this chapter of the tension MDBs face in pursuing their banking objectives versus their development objectives. Korten, for example, studying the negative impact of an Asian Development Bank loan for Philippines forestry projects, argued that the banking prerogatives can by themselves be environmentally destructive. Loans made by MDBs must be repaid in foreign exchange, which can create pressure for recipient countries to export natural resources.[68]

Wade's work, as mentioned above, describes numerous examples of other performance and budgetary systems at work, even in the midst of the World Bank's most concerted efforts to improve its environmental management. These include instances of poor cooperation between the Bank's regional environmental departments and its operational departments, weak project management during times of Bank-wide reorganization, staff ability to ignore quality control mechanisms, and disincentives for environmental specialists to be "tough" in addressing environmental aspects of projects. His causal arguments point to several factors, including a clash of highly different worldviews among bank staff, NGOs from donor and borrowing countries and other actors interested in the Bank's services, and the tactical nature of many of the Bank's reform efforts in response to outside pressure.[69]

Within the broader social scientific literature, several strands of study highlight the impact of institutional mechanisms on institutional behavior. Historical institutionalists, for example, are concerned with the ways in which preferences shape institutions and vice versa.[70] This focus has been most evident in scholarship on European integration, particularly among scholars exploring the relationship between European institutions and domestic policy and the conditions under which institutions pursue their own preferences instead of follow the wishes of their shareholding countries.[71] In one incisive attempt to show when and how each actor matters, Pierson argues that realist perspectives can explain political bargains at *specific points in time,* while an analysis of evolving processes reveals that institutions can function in ways not anticipated by their designers.[72] While member states are still the central decision makers in the process of European integration, their control is limited by a number of factors that create gaps in member state control, and which give other actors the ability to influence the integration process. These gaps are caused by factors such as: member states must allow EU organizations to have a degree of autonomy to complete their increasingly complex tasks; decision makers have short time horizons, which means that they often do not anticipate the long-term consequences of institutional design and reform; and finally, member state policy preferences shift over time, which can result in the institutional arrangements diverging from member states' original intentions. Pollack, using a principal-agent framework, argues that EU institutions are characterized by varying patterns of autonomy, since the efficacy and credibility of member state control mechanisms vary across institution, issue area, and over time.[73] Many of these observations resonate with the case of MDBs, particularly in their assumption that institutional behavior is shaped by different exogenous political and internal institutional factors at different stages of the policy process. The argument presented here adds to the analysis of how these mechanisms work when institutions are juggling potentially conflicting mandates, as well as how similar mechanisms function across similar types of institutions.[74]

Branches of public choice institutionalism and rational choice institutionalism also analyze what happens inside institutions. Shepsle, for example, shows how frameworks of rules, procedures, and arrangements constrain and mediate individual preferences to shape outcomes.[75] Sociological organization theories and theories of bureaucratic politics also look inside the black box of the organization to explain behavior.[76] Rational choice arguments tend to differ from the latter group in their underlying assumptions that institutions reflect aggregations of individual choice, versus the sociologists' view that institutions are embedded in cognitive, cultural, or political foundations. They also differ in their views on whether preferences are fixed and exogenous, or as March and Simon have argued, rationality is "bounded" by a lack of information, a great deal of uncertainty, and in general, people's limited ability to process information and solve problems.[77]

The vast majority of work by organizational theorists focuses on domestic government organizations or firms, rather than IOs. Yet international relations scholars in recent years have begun to examine whether and how lessons and causal variables offered by organizational theory might apply to the study of international institutional behavior. Interestingly, much of this work has been prompted by growing analytical attention to the way international institutions have responded to global environmental issues.

In his analysis of global environmental governance, for example, Oran Young has argued that it is important to look at decision procedures, compliance mechanisms, and sources of revenue in determining the effectiveness of international institutions.[78] He contrasts these endogenous variables with exogenous variables that include the broad array of "physical, biological and social" conditions that make up the environment in which IOs operate. Some of the case studies highlighted in an edited volume by Keohane and Levy focus on the ways institutional design variables govern the extent to which financial mechanisms for environmental aid have either been effective or—more often—ineffective in promoting environmental changes in poorer countries. Within this project, Connolly, Gutner, and Bedarff have shown how environmental aid by the largest Western donors to Central and Eastern Europe in the early 1990s followed Cohen, March and Olsen's "garbage can" model, where new environmental protection goals were tacked onto preexisting institutional objectives. Solutions did not necessarily fit the problems, and the result was the neglect of certain priority problems in the region, duplication of assistance efforts, and failures in aid coordination.[79]

Ultimately, this area of research requires more empirical work to test and compare which design- and rule-related variables have the greatest impact on institutional behavior, and to determine how they do so. In the case of the MDBs, chapters 4 and 5 will show how variation in how banklike or demand-driven an MDB is designed to be has an important impact on staff incentives, capacity, and resources. For example, a less banklike MDB will have more "green" bankers, people whose job is to actively search out and design environmentally oriented projects. Where environmental staff are responsible for identifying environmental loans and designing projects at the early stages—that is, where they essentially act as environmentally oriented bankers—the MDB will produce more stand-alone environmental projects. Where there is a separate environmental appraisal unit with the power to change the shape of a loan as it is being designed, the MDB will be more proactive in its environmental work.

Recipient Country Incentives
Curiously, analyses of MDB environmental behavior do not pay much attention to incentives from the perspective of the recipient country; that is, loan recipients may have little or no interest in the environmental components of loans, or stand-alone environmental projects, and may thus block the bank's ability to implement its environmental policies. Analysts occasionally touch on the fact that recipient country political support may affect what is perceived as the bank's environmental behavior. For example, Rich and Aufderheide noted: "Even (World Bank) bureaucrats with the best intentions often have their hands tied. The policies they implement are frequently shaped by political pressures that these functionaries are generally powerless to confront or alter. . . . Even well-intentioned forestry

experts within the World Bank may be thwarted in their efforts by their counterparts in recipient countries, since national forestry agencies are often staffed by individuals who profit from deforestation." [80]

George and Sabelli have also noted that the World Bank often has a difficult time convincing recipient countries to accept environmental conditionality. "Borrower governments," they wrote, "read environmental protection clauses in loan agreements as code for 'this project is going to cost us a lot more and will include completely superfluous but mandatory components we'll have to pay for.' " [81] Indeed, ministers of forty-one developing countries declared their opposition to environmental conditionality in a June 1991 preparatory meeting ahead of the 1992 UNCED conference. [82]

These issues are not well explored in the policy literature on MDBs and the environment, most likely because many of the author-advocates are more intent on producing changes at the bank level, with a focus on institutional design, governance, and performance issues. In other words, many of the policy advocates focus on the *supply* of environmental projects and policies by MDBs rather than the *demand* for such projects and policies. However, the interests of recipient countries and their impact on the ability of IOs to carry out their work is addressed in the broader policy literature on MDB conditionality and aid effectiveness, which provides some hypotheses examined in this study.

Studies focusing on the implementation of donor policies and projects in recipient countries highlight three key variables that appear to explain how incentives at the recipient's level influence the effectiveness of project implementation. First, and unsurprising, is the argument that donor policy goals are unlikely to be met unless they intersect with the recipient's own priorities. [83] This can occur either through the existence of a set of mutually agreeable policy goals, or more likely as a result of bargaining between the two sides. Mosley, Harrigan, and Toye, analyzing the efficacy of World Bank policy-based lending (or structural adjustment loans), show how conditionality is a bargaining game, where the donors are interested in seeing their policy choices carried out, while the recipients resist any conditions that differ from their own policy preferences. [84] The outcome of bargaining between the two cannot be predetermined; it depends on the actors' strategies, as well as the external environment. Recipients or donors may have stronger bargaining positions in some cases than in others. Studies of project effectiveness also conclude that aid conditionality is most likely to make inroads in cases where it strengthens domestic interest groups, or the political authority of officials committed to the same goals.[85]

Since economic restructuring has been a primary goal of Central and Eastern European governments, we would expect to see donors making more inroads on environmental issues in their public sector lending where the policy changes desired also contribute to economic restructuring. In other words, recipients will be more accepting of projects or policies with environmental components where these projects also have important economic benefits. A key example of this is the call by most MDBs for governments to reduce subsidies on energy prices. The removal of subsidies is an important step toward economic liberalization and away from price distortion. On the environmental side, it would reduce air pollution by reducing demand for energy, and also would encourage more efficient use of energy. Reducing subsidies on coal, the primary source of energy for many countries in CEE, also increases the attractiveness of alternative, cleaner sources of energy, such as gas. By the late 1990s, those CEE countries preparing to join the EU had fresh, powerful incentives to pursue investments in the major environmental sectors (water supply and waste water, air and waste management) in order to approximate European environmental laws.

In the case of MDB activity in CEE, what is interesting is that the set of MDB projects that benefit the economy and the environment also happen to be the same set of projects that do not necessarily require new behavior from the banks. These are activities where the banks' economic and environmental mandates may already overlap. The lessons from these observations focus attention on the need to analyze the MDBs' portfolios to determine: (1) if they have evolved in the ways loans are designed in areas where traditional economic criteria overlap with new environmental criteria in order to better address relevant environmental issues; and (2) if they attempt to push *beyond* the joint set of activities most preferred by recipients on economic grounds, as a way to better tackle local, regional, or global environmental problems.

The second variable of key importance in determining project success is the administrative capacity of the recipient at the national, subnational, or even firm level (in the case of private sector loans), to help carry out implementation.[86] The best-designed projects will fail if the recipient does not have the ability to help implement them and maintain them, once funding ceases.[87] This variable is not likely to point to significant variation among the banks. Projects from all three banks are likely to suffer more in countries with weak administrative capacity, and have a greater chance of success in countries with stronger domestic capacity.

The third important variable in determining effective project implementation is the ability of donors to manage potential principal-agent problems to ensure that implementing agents carry out projects as planned. This refers to the accountability system set up by the MDBs, which, in addition to rules embodied in loan covenants, includes systems of reporting, monitoring, and evaluating projects. MDB projects can be compared at this level to see the extent to which the banks follow through with their work once the loan agreements have been signed and money has been disbursed. Such comparison is important in revealing the degree of "slippage" that occurs in the implementation of bank projects.

Goal Congestion

Some analysts argue that the key factor inhibiting the World Bank from being more effective in fulfilling its policy goals is the fact that the sheer number of goals has mushroomed to a level that makes it virtually impossible for the institution to meet them all, which results in what has been called "goal congestion" and "mission creep."[88] This proliferation is the result of a diverseness of views among donor countries, recipient countries, and bank staff and management about what the primary priorities of the institution should be, along with the Bank's inability to shed old goals as it moves to adopt new ones. As Naim noted: "In the case of the World Bank, the lack of consensus about its basic mission, limitations in its governance system, and other conditions have led to a proliferation of goals—which in turn has had important organizational repercussions. . . . Organizations like the Bank—large, complex, relatively autonomous, and with a significant capacity to influence its environment—can postpone, or even avoid, the difficult decisions required to minimize incongruities between strategy and internal organization."[89]

From a slightly different angle, other analysts argue that it is not necessarily the number of competing ideas and proliferating goals that matters, but rather the existence of too many parties to which the Bank must be accountable. Horberry argues that international agencies are accountable to too many actors—donor governments, recipient governments, funding sources, and other actors involved in the process of planning and implementing development aid. In the face of too many demands, which sometimes conflict, the agencies cannot satisfy all parties, and "tend to pay lip service to the general objectives of new policies without dramatically changing the actual management of their programs."[90]

In some ways, this argument directly challenges the "weak governance" explanation. It implies that the institution is struggling to be accountable to a plethora of policies and sources, while the latter argument implies that the institution chooses not to be accountable to its owners. Yet, the outcome of both arguments is the same; the result of either source of ambiguity increases the autonomy of the MDB's staff and its ability to shape the bank's strategies and policies.[91]

Curiously, this vein of argument has no direct parallels in the broader institutional literatures, although it clearly bolsters arguments by neoliberal and comparativist scholars who stress the importance of path dependency and unintended consequences. It is an argument that requires greater exploration through empirical research. On one hand, it appears to have little explanatory power vis-à-vis the World Bank's environmental behavior in terms of portfolio development. This is because the World Bank clearly has more goals and mandates to juggle than the other two MDBs, and yet this has not appeared to hurt the World Bank relative to the others in its planned environmental activities in Central and Eastern Europe. Indeed, despite its higher degree of "goal congestion," the World Bank appears to be significantly more active than the other two in marrying environmental and economic project and policy goals. On the other hand, the plethora of policy goals facing the Bank appears to be a factor that makes complex projects difficult to implement. In other words, although goal congestion has not stopped the World Bank from engaging in a number of environmental project and policy activities in CEE, it may contribute to a gap between the environmental considerations built into project designs and the extent to which they are carried out in practice, as well as the speed at which projects can be implemented. In analyzing MDB behavior, it is important to distinguish between problems that may be caused by a *congestion* of goals, and problems caused by *conflicting* objectives, since the two may not always be the same.

Conclusion

This chapter has shown the ways in which the MDB-centric literature and the broader scholarly literature intersect and also diverge in their emphases and arguments. More importantly, in examining the explanations offered by each, I argue for the importance of linking explanations that seek to explain how donor preferences and institutional variables interact in explaining institutional behavior. Single variable explanations may explain a piece of the picture, but are not sufficient to address the complexity of how donor politics and institutional design and incentives shape MDB behavior. This argument is expanded in subsequent chapters.

3

Bargaining and Delegation: The Birth of Environmental Mandates

Multilateral development banks, traditionally most comfortable making loans to creditworthy counterparts for infrastructure-related projects combining the expertise of economists and engineers, are now going to great lengths to show how "green" they have become. In recent years, attention to environmental issues has become an important goal for the World Bank, the EBRD, and the EIB—at least in rhetoric. The changes are most evident at the World Bank, where there has been an explosion of environmental policy activity and lending over the past decade. The EBRD, the youngest of the three, has an environmental mandate built into its charter, which requires it to promote "environmentally sound and sustainable development" in its activities.[1] The EIB, in turn, points out that financing environmental projects is "a major activity of the Bank," and that "the environmental issues of project financing" are "a major concern."[2] The three banks share the same two basic environmental objectives: they seek to address and mitigate potentially adverse environmental impacts in their projects, and they seek to fund projects with major environmental components or goals. Yet, the ways they define these goals and seek to carry them out differ considerably.

This chapter addresses the first stage of the policy process. It introduces the three banks and examines how and why they decided to take on new environmental policy goals. An analysis of the political and ideational forces behind the banks' adoption of explicit environmental goals reveals the strength of each bank's apparent commitment to environmental issues. Pressure from major shareholder countries, usually supported or pushed by environmental NGOs, is a key factor in determining how seriously an MDB will decide to address environmental issues in its work and explains why the World Bank and EBRD are "greener" than the EIB.

Although environmental policy goals can be generated by bank actors, they tend to languish without external support or pressure.

Where there is evidence of major shareholder pressure for change, it has been prompted by pressing problems, or what Kingdon calls "focusing events" that capture the attention of major shareholders. Analyzing agenda-setting processes in the U.S. federal government, Kingdon argues that important policy changes occur when independent process streams of problems, policy options, and politics join together at a "critical juncture," most often when a "policy window" opens owing to a compelling problem or changes in political factors such as interest group pressure.[3] For the World Bank, the existence of several projects publicized by NGOs as environmental disasters gave these organizations ammunition to successfully heighten Congressional awareness of the Bank's weak environmental accountability and unleash U.S. pressure on the Bank to revise its environmental goals. In the case of the EBRD, donor awareness of the enormous environmental problems facing former communist countries seeking to embark on economic reform put environmental issues high on the agenda of those negotiating the new Bank's articles of agreement. Again, U.S. influence, backed by NGO support, was quite visible. For the EIB, the absence of such a "focusing event" is one important reason for the absence of strong shareholder pressure on the Bank to be more ambitious in its environmental objectives. The absence of the environmentally activist United States on the EIB's board is also a reason for that bank's relatively quiet environmental stance.

The World Bank

Basic Mission and Lending Policy

With its lending peaking at almost $30 billion in fiscal 1999, the World Bank is the biggest single creditor for developing countries and countries in transition. It is the oldest MDB, and the only one with a truly global scope.[4] With an expert staff of just over 9,000, it is a major source of policy advice and technical assistance for borrowers. The Bank and its sister institution, the International Monetary Fund (IMF), were born in 1944 at Bretton Woods, New Hampshire, when representatives of 44 countries met to lay the foundation for the postwar economic order.[5]

The Bretton Woods institutions, which opened their doors in 1946, were created to promote economic cooperation, growth, and stability in an effort to avoid any return to the beggar-thy-neighbor trade practices from the prewar years. Both multilateral institutions would be owned and run by national governments, but their governance was clearly dominated by the United States. The United States both proposed the initial concept of the World Bank and played a leading role in its design, policies, and tools.[6] With the United States providing most of the money to start the Bank, its delegation was also able to ensure that the Bank would be located in Washington, D.C. From the beginning, the president of the World Bank has been American, while the head of the IMF has been European.

According to the World Bank's Articles of Agreement, it would operate by guaranteeing or making loans for projects for reconstruction or development. In practice, the Bank focused on direct government lending, backed by sovereign guarantees, since governments preferred to borrow directly from the Bank rather than guarantee private sector lending by the Bank. Voting power of the Bank's shareholder governments was determined by the size of each country's capital subscription to the Bank. The United States began with a 34.9 percent stake in the Bank.[7] Although that stake has declined over the years (reaching 16.98 percent by fiscal 1999), the United States has remained the Bank's single largest shareholder and most influential board member.[8] Beginning with authorized capital of $10 billion, out of which members subscribed $7.5 billion, the Bank committed itself to conducting its activities "with due attention to considerations of economy and efficiency and without regard to political or other non-economic influences or considerations."[9] Loans and guarantees would be funded by the shareholders' capital subscriptions, plus principal and interest payments on loans, and later (and most important), funds raised from international capital markets.

The World Bank is no stranger to shifts in its institutional goals and missions, and over the years the ways it has chosen to define and promote development have evolved considerably, in response to trends in thinking, to criticism, and to a broader changing global context in which it works. Its formal name, the International Bank for Reconstruction and Development (IBRD), reflected the dual-emphasis desired by its founders. In the

Bank's first year or so, the financing of postwar reconstruction was its primary function, but any possible tensions between the two priorities were soon displaced by the Marshall Plan's grant-dominated funding, which sharply reduced Western Europe's demand for World Bank loans.[10] The Bank's early emphasis was on promoting big, public sector, capital infrastructure projects, such as railways, roads, ports, power plants, and communications equipment, as a way to create a foundation for private sector investment, and hence economic growth. As Mason and Asher note in their seminal history, the early years were characterized by a narrow conception of development and by an overall "lack of attention to development analysis."[11]

In the 1950s and '60s, as the World Bank's focus shifted toward addressing the challenges facing developing countries, its view of development widened, the methods it used to achieve its goal of promoting economic development evolved, and new affiliates were created. The latter included the International Finance Corporation (IFC), established in 1956 as the Bank's private sector lending arm, with the ability to lend without government guarantees, and the International Development Association (IDA), established four years later to provide highly concessional financial resources to poor countries unable to qualify for regular Bank loans.[12] The Bank began to extend its activities in agriculture, industry and even in education lending. During Robert McNamara's thirteen-year tenure (1968–1981), the Bank's central mission shifted from "closing the gap between industrial and developing countries to alleviating world poverty," and its staff and lending volume increased sharply and included expanded programs for rural and urban development.[13]

In the early 1980s, among other changes, the World Bank embarked on the controversial nonproject, structural adjustment lending (SAL), or lending to support macroeconomic policy changes. While poverty alleviation was restated as a central objective under James Wolfensohn's tenure beginning in 1995, the list of specific lending objectives has continued to expand over the years, adding a range of new issues that includes work on promoting opportunities for women, strategies to fight HIV/AIDS, reducing corruption in recipient countries, and of course, improving environmental management.[14] In addition to lending money for this broad array of issues, the Bank is a major source of policy advice to borrowing countries. More than any other MDB, the World Bank has an intellectual life, bolstered by a vast research arm supported by a multimillion dollar budget.[15] Lending instruments also include investment loans for particular sectors, adaptable program loans for long-term development programs, loans for technical assistance on a variety of policy reform issues, and sectoral adjustment loans, among others. The Bank's range of lending and its nonlending activities define it as the least banklike of the three MDBs examined in this book, an issue we will return to in chapter 4.

Birth and Evolution of the Environmental Initiative

There are two distinct periods in the history of the World Bank's environmental thinking. The first period, 1970–1987, marks the birth of an explicit environmental initiative. This period established the Bank as a pioneer among multilateral institutions in thinking about how to address environmental issues in its work. However, weak institutionalization of the initiative meant that the rhetoric far outweighed the results. The second period, 1987 to the present, has been a period of significant change in the ways in which the Bank has acted on environmental issues in terms of its internal policies, organizational structure, and staffing. In addition to its portfolio of stand-alone environmental projects, which totaled around $6 billion through fiscal 1999,[16] the Bank was also the leading implementing agency of the Global Environmental Facility and the Montreal Protocol's Multilateral Fund, through which it funds projects promoting biodiversity, renewable energy projects, and investments to help the phaseout of ozone-depleting substances, among other areas. The story of the evolution of the Bank's environmental policy goals owes much to significant United States pressure in the late 1980s—from Congress and the Treasury, which was strongly aided and pushed by a group of environmental NGOs.

1970–1987   The World Bank's first explicit foray into addressing environmental issues was launched in 1970, when Robert McNamara announced that development institutions faced the challenge of helping developing countries "avoid or mitigate some of the damage economic development can do to the environment, without at the same time slowing down the pace of economic progress."[17] He set up a new unit in the Bank to address this issue: the Office of Environmental and Health Affairs, which soon became the Office of Environmental Affairs (OEA).[18]

A few important factors account for the timing of this decision. The 1960s gave birth to active environmental movements in the West, reflecting a sharp increase in popular concern about environment issues.[19] In the early 1970s, countries around the world were preparing for the first United Nations theme conference, the 1972 United Nations Conference on the Human Environment (UNCHE, or the Stockholm Conference), which publicized global environmental issues and gave birth to the United Nations Environment Programme (UNEP).[20] The conference's main organizer, Maurice Strong, called on the World Bank to help define the issues that would be discussed at the conference. In fact, the "Declaration, Principles, and Recommendations" produced by the conference was based on an early document (the Founex Report) that was drafted mostly by World Bank officials.[21] McNamara himself gave the conference's keynote speech.

Other intellectual and political factors also shaped McNamara's decision. Le Prestre and others have pointed out the influence of McNamara's close friend, Barbara Ward, who pushed for the World Bank to be active in addressing environmental issues.[22] Ward, who went on to establish the International Institute for Environment and Development, also cowrote a "conceptual framework" commissioned for the Stockholm conference, which later became a bestseller.[23]

Finally, there was greater interest within the United States for deeper examination of environmental issues, which translated into support for the Bank to take more action on this front. In 1969, the United States Congress passed the National Environmental Policy Act, which obligated federal agencies to prepare environmental impact statements on proposed programs that might have a significant impact on the environment.[24] As Le Prestre noted, the World Bank closely watches the actions of its largest shareholder. At the same time, there was no great show of environmental support from the other major executive directors.[25]

Although OEA's creation was an important move at the time, its performance was lackluster. The initial momentum behind its creation fizzled out in subsequent years, while politically, there were few long-term supporters. The United States, for example, turned its attention to other issues during the 1970s, notably the oil crisis, while its relations with the World Bank soured over a number of disputes.[26] Recipient countries, in turn, did not include environmental improvement among their list of priorities and generally viewed environmental protection as a luxury for developed countries to address. Until the mid-1980s, NGOs were not paying close attention to the Bank's environmental behavior. Bank staff, meanwhile, viewed the OEA skeptically and were resistant to the OEA's scrutiny of Bank projects.

Ultimately, McNamara's policy decision to strengthen the World Bank's environmental scrutiny was undermined by the fact that OEA lacked the power or resources to do much. The job of the environment office was to screen Bank projects with an eye toward reducing potentially harmful environmental affects. The environmental team mainly saw projects in the last stages of the project preparation cycle, which reduced its ability to introduce significant modifications. The OEA had a tiny staff—only three specialists by 1983, and five by 1987—at a time when Bank lending had grown to over 200 projects a year. It also had no veto power to block projects from moving forward for board approval. The Bank did issue a set of environmental guidelines in 1984, but these were easily ignored, since they left compliance to the discretion of the task managers.[27]

Moreover, during this period, the World Bank's understanding of what constituted an environmental issue was never explicitly defined. By late in Alden Clausen's presidency (1981–1986), "central" environmental issues included everything from pollution to the preservation of "mankind's aesthetic and cultural heritage."[28] The environmental office was envisioned to be a trouble-shooter, a group that would try to fix potentially problematic project components. In practice, its emphasis tended to be on public health issues, and more specifically disease prevention, such as water-borne disease resulting from irrigation and hydroelectric projects. This was the area of expertise of the Bank's Environmental Advisor, James Lee.[29] There was also attention to "brown" issues, such as industrial pollution, as well as the use of pesticides.[30] As Wade concluded: ". . . what the Bank did under the label 'the environment' included residual things like the relocating of a power line so as not to spoil the view from a game lodge, matters that no one else wanted to deal with."[31]

Ultimately, the lack of strong political initiative from shareholders, plus weak institutional incentives inside the World Bank, thwarted the possibility of serious attention to or definition of environmental issues throughout the Bank during this period.

1987–2000   In 1987, the World Bank launched a process of significant change in the ways it would address environmental issues when World Bank president Barber Conable admitted the Bank had in fact been "part of the (environmental) problem" and announced an important set of organizational changes. In a speech to the World Resources Institute in May 1987, Conable admitted that "in measuring the influence of the World Bank against the environmental challenge, I see how long a road there is to travel from awakened environmental consciousness to effective environmental action."[32]

He announced the creation of a central environmental department, which would develop environmental policies, planning, and research, and "take the lead in developing strategies to integrate environmental considerations into our overall lending and policy activities." In addition, he announced that new environmental offices would be established in the Bank's four regional technical departments (called regional environmental divisions, or REDs), which would work as "environmental watchdogs" over the Bank's projects.[33] The REDs would have the ability to review and sign off on projects, which allowed them to influence lending and created what Wade calls "in-house environmental champions."[34] By the end of the year, around fifty new environmental jobs had been created in the Bank.

The years following Conable's announcement have seen extensive institutional and policy changes in terms of how the World Bank considers and acts on environmental issues. The Bank vastly expanded its work in developing environmental policies and procedures for all projects as a means of reducing potential environmental harm. It sharply increased its research on environmental issues and practices and created a new vice-presidency for environmentally sustainable development, which is the umbrella over the Environment Department, the Transport, Urban, and Water Department, and the Agriculture and Natural Resources Department. It developed sectoral strategies and assisted recipient countries in the development of national strategies (known as national environmental action plans, or NEAPs), and finally, it has emerged as a global and regional leader in a number of multinational environmental initiatives. The latter includes its role as an implementing agency for the GEF and the Montreal Protocol Multilateral Fund, and lead organizer and author of the Environmental Action Programme for Central and Eastern Europe

(EAP), endorsed by environmental ministers from East and West in 1993. According to the Bank, around 300 Bank staff and long-term consultants work on environmental issues, but this number is more of a rough guideline than a precise figure.[35] On the surface, at least, the Bank by the early twenty-first century looked quite different than it did in the late 1980s.

The numerous changes initiated in 1987 were triggered by external criticism of the Bank by environmental NGOs coupled with strong United States pressure for change from Congress and the United States Treasury. An NGO campaign to force the Bank to change its environmental ways took its ammunition from the existence of a handful of Bank projects that were causing enormous environmental degradation (particularly the Polonoroeste project and also the Indonesian Transmigration Program).[36] The campaign effectively built support within Congress at a time when the United States was negotiating over the size of its IDA contribution.

Capital replenishments have become a key way for shareholders and NGOs to lobby for policy changes within MDBs. In the case of the World Bank, the fact that IDA requires regular replenishments has given its more active shareholders, primarily the U.S. Congress, the ability to use the power of the purse strings to influence Bank policy.[37] NGOs and other interest groups can focus their lobbying efforts on the various congressional committees involved in U.S. policy toward the World Bank.[38]

The NGO campaign for environmental reform at the World Bank was launched by the Natural Resources Defense Fund (NRDC), the Environmental Policy Institute (which later merged with Friends of the Earth), and the National Wildlife Federation (NWF); many other groups quickly joined in. These NGOs began investigating the Bank's environmental behavior in 1983 and were deeply unimpressed with their findings. Bruce Rich, then at the NRDC, was one of campaign's leaders. He was highly critical of the ability of the Bank's environmental office's small staff to evaluate projects effectively, and believed that the Bank's procedures and guidelines were essentially worthless:

The Bank vaunted its publicly available environmental guidelines, checklists, and procedures, bound in two thick blue volumes. Indeed scores of dust-covered sets sat in piles inside the offices of the environmental staff, available to anyone from the outside for the asking. In an institution that zealously kept its most trivial working documents secret, this was no anomaly: not only were the Bank's other 6,000 employees not required to use the guidelines in preparing projects, most of them were not aware or their existence. Indeed, it would have been senseless to require most Bank staff to use them, since the guidelines focused on industrial and processing activities, a small part of Bank lending in comparison to agriculture, irrigation, dams, and roads—for which there were no guidelines whatsoever.[39]

The environmentalists chose the World Bank as the focus of their campaign not necessarily because its environmental record was worse than other MDBs, but because it was the best known, it was based in their own city of Washington, D.C., and because they felt they could influence the Bank's policies through Congress.[40]

In 1983, the NGOs convinced the House Banking, Finance, and Urban Affairs Subcommittee on International Development Institutions and Finance to hold oversight hearings on MDBs. These were the first of over twenty hearings on MDB performance held before six Congressional subcommittees between 1983 and 1987.[41] The NGOs brought in their own advocates, consultants who had worked on disastrous projects, and representatives of indigenous people's rights organizations, who detailed the degree of social and environmental harm caused by some of the Bank's projects. Congress reacted by developing, with NGO assistance, a number of recommendations for environmental reform of the MDBs. These recommendations directed the United States to urge MDBs to increase their environmental staffs, to consult with NGOs and environmental and health ministries in project preparation, and to fund more small, environmentally beneficial projects.

By 1986, the campaign had grown, with a number of European and developing country NGOs involved, as well as fresh support from other major shareholders, including Germany, Sweden, the Netherlands, and Austria.[42] As important, the U.S. Treasury, the World Bank's official U.S. shareholder, began increasing pressure on the Bank to reform its environmental practices. In 1987, Congress would decide on its participation in IDA's eighth replenishment, as well as a more rare IBRD capital replenishment.[43] Given the degree of Congress's frustration with the Bank's environmental behavior, the Treasury realized that an increase in the U.S. contribution to IBRD or IDA would be difficult in the absence of environmental reform. "There was no doubt," wrote Rich, "that the environment had become the Bank's most prickly public relations problem."[44]

Barber Conable, who became president of the Bank in July 1986, had been a Republican congressman for more than twenty years. He was sympathetic to the environmentalists' concerns and well understood the obstacles faced in Congress to increased U.S. funding. The result was his May 1987 speech, announcing a set of environmental reforms for the Bank.

Since 1987, the Bank's environmental approach has continued to evolve, with changes reflecting policy development by staff, but also several initiatives that can be traced to external pressure and shareholder initiative. During the IDA-9 replenishment negotiations in 1988, Congress pushed the Bank, through the Treasury, to propose environmental impact assessment (EA) procedures. In this venture, they had support from some Bank officials, which included two experts on environmental impact assessment within the Bank's REDs. These men had developed EA procedures for their respective regions and enlisted the help of a third colleague to work on a Bank-wide policy. In 1989, some twenty-six drafts later, an operational directive on environmental assessment was completed, despite clear opposition from a number of Bank officials. Conable, the U.S. Executive Director, and some executive directors from other main shareholders supported the changes.

Two other environmental initiatives stemming from shareholders are the GEF and the Montreal Protocol's Multilateral Fund. In 1989, developed country governments asked the World Bank to design what would become the GEF, which would provide grants to help poor countries address four global environmental problems: global warming, biodiversity, ozone layer depletion, and the pollution of international waters. The Bank is the main implementing agency of the GEF (its partners are UNDP and UNEP). Donor countries also asked the Bank, along with UNDP and UNEP, to manage the Montreal Protocol's Multilateral Fund, which finances the incremental costs developing countries face in phasing out ozone-depleting substances under the Montreal Protocol.[45]

A third important initiative worth noting is the World Bank's Inspection Panel, established in 1993. This three-member commission is charged with investigating complaints against Bank activities by private entities and groups who believe the Bank has not followed its own policies and procedures. Although this evolution in Bank policy is not directly linked to environmental issues per se, criticisms of the Bank's lack of accountability did stem, in part, from environmental issues and groups. Once again, the United States was a leading shareholder pushing for change, with support from several other shareholders; in addition, U.S. NGOs lobbied strongly for change. Again, negotiations about IDA replenishments helped donors supporting the creation of an independent panel to win their case. In this case, United States NGOs urged Congress to withhold a portion of United States funding for IDA-10 unless the Bank created some type of commission to hear complaints against the Bank's work. A modified version of the NGO proposal entered U.S. appropriations legislation.[46] The accountability issue was clearly of importance to the United States, which was also pushing the Bank to adopt a new information disclosure policy, instituted in 1994.[47]

General Environmental Policy Objectives

By the late 1990s, the Bank's main environmental goals had evolved in a number of ways. It emphasized the development and use of "safeguard" procedures and policies to ensure that its projects are not environmentally harmful. It sought to fund more projects with primary or significant environmental components, and to assist countries in strengthening their institutions, strategies, and policies for environmental management. The Bank also began promoting the idea of the importance of "mainstreaming" environmental issues into its broader array of projects and programs.

Of the three MDBs, the World Bank has the most comprehensive set of policy goals, and by far, the most literature explaining these goals.[48] Its work on minimizing potentially negative environmental impacts of its work has expanded beyond its EA procedures to include the development of policies for sectoral lending in forestry, agriculture, water resources, energy and transport, pest management, and natural habitats, among other areas. It often takes the lead in initiating national and regional environmental action programs (NEAPs or REAPs) that seek to prioritize environmental problems, discuss policy options for solving them, and increasingly, link these studies to specific investments. The Bank has also sought to incorporate environmental issues into its country policy work and country assistance strategies that guide lending, as a means of helping recipient countries develop programs and institutions for more sound environmental management. In addition, the World Bank has undertaken a significant amount of research on a broad scope of environmental issues, from specific natural resource issucs to the complex interaction between environment, economic growth, and poverty.

External political pressure therefore has given the World Bank a strong commitment to address environmental issues in its work, but as we will see in subsequent chapters, this commitment is not easily transformed into deep changes in the incentive systems facing Bank staff and management, which has resulted in a mixed performance. The World Bank's challenge has been how to translate its policies and research into environmental performance, through its project and policy work.

EBRD

Basic Mission and Lending Policy

The London-based EBRD is the first post–Cold War international institution, established in May 1990 with 10 billion ECU (around $13 billion) in capital. It was a European initiative created to help postcommunist countries in Central and Eastern Europe build market economies and pluralist democracies.[49] The proposal was put forward by French President François Mitterand at the European Parliament in October 1989 and was discussed by European Community leaders in Paris and Strasbourg meetings over the next two months.[50] As communist regimes began collapsing across the East, there was a growing sense among major Western nations that aid and capital flows to CEE needed to be dramatically increased to assist the nascent democracies with their unprecedented process of economic and political transition. The French proposal was fleshed out by Mitterand's advisor, Jacques Attali, who envisioned an institution with a limited role for non-European countries. European leaders meeting in Strasbourg modified the French proposal to include a larger role for the United States, Japan, Canada, and other non-European countries.

A number of countries, including the United States, West Germany, Italy, and the Netherlands were skeptical of the need to create a new institution.[51] From the United States' perspective, the World Bank and the IFC already financed the types of activities envisioned for the new bank. U.S. officials were also unenthusiastic about possible Soviet participation in a new international institution, with a number of conservative

Republicans arguing that the Soviet Union was not yet a democracy working toward creating a market economy.[52] Some European as well as U.S. officials, in turn, observed that the EIB could also perform the activities envisioned for the EBRD. The counterarguments included French pressure for a specifically European institution, and one that was more politically aggressive than the EIB. A European-led institution would not only signal Western Europe's support for its Eastern neighbors, but would also play a role in diluting the prospects of German economic dominance in the former East Bloc.[53] Those backing the creation of a new institution also argued that Central Eastern European and former Soviet countries could not sit on the board of the EIB, since the latter was exclusively a European Community institution.[54] The EIB, in turn, was distinctly uninterested in accepting responsibility to lead a Western aid initiative in CEE.[55]

After several more negotiating meetings over subsequent months, on May 29, 1990, the negotiating governments, along with the European Community (EC) and the EIB, signed agreements establishing the EBRD. It was remarkable that a new international institution with billions of dollars in capital could be proposed and created in less than a year. It took a greater period of time—an additional eleven months—for the articles to come into force, and the new Bank opened its doors in April 1991.

The EBRD's mission was to "foster the transition towards open market-economies and to promote private and entrepreneurial initiative in the Central and Eastern European countries committed to and applying the principles of multiparty democracy, pluralism, and market economics."[56] It sought to fulfill its mission by emphasizing projects in privatization and economic restructuring, by supporting foreign investment (in part by investing in joint ventures with foreign companies), by cofinancing with the other MDBs, and by promoting small and medium-sized enterprises.

The Bank was distinctive in having an explicit political mandate, emphasizing its commitment to the principles of multiparty democracy, market economies, human rights, and the rule of law. This was in sharp contrast to charters of other MDBs, which prohibit them from being influenced by political decisions. And unlike the World Bank and EIB, it was set up with a focus on the private sector. The Bank was directed to target at least 60 percent of its loans, guarantees, and equity investments to the private sector, with the remaining 40 percent going to public sector projects. This gave the Bank the character of being part merchant bank and part development bank. With its emphasis on the birth and growth of the region's private sectors, it saw itself having a more decentralized, "bottom-up" approach, versus the World Bank's "top-down" public lending approach. Finally, the EBRD was the first MDB to have an environmental mandate built into its Articles of Agreement; Article 2(1)(vii) commits the Bank "to promote in the full range of its activities environmentally sound and sustainable development."[57]

Like that of the World Bank, EBRD membership was structured so that a country's shareholding stake translated into its voting power. A notable difference between the two is that in addition to member countries, the EBRD's board also included two *institutions:* the European Commission and the EIB. (Likewise, the Commission is on the board of the EIB.) The United States was given the single largest stake, of 10 percent, while the twelve EC nations, as well as the Commission and EIB, held a majority of 51 percent of the votes. The Soviet Union was allowed to join under the proviso that it did not borrow more than its contribution to the Bank's capital—a small $125 million—but this restriction was gradually eased once the Soviet Union collapsed soon after the Bank's inauguration. The Bank's "countries of operation" held around 12 percent of the shares. Jacques Attali, one of the key intellectual forces arguing for the creation of the Bank, was appointed its first president. By 1999, the Bank's members included 58 countries (of which 26 were recipient countries in CEE and the former Soviet Union), as well as the two EU institutions. Germany, France, the U.K., Japan, and Italy are the next largest shareholders after the United States, with 8.5 percent each.

In the Bank's relatively short life span, its central mission has not changed as dramatically as the World Bank's did during its fifty-plus years of existence. Indeed, the EBRD's main struggle during the first few years was to jumpstart its operations—no easy task when its staff sought to find private sector projects to fund in countries where no private sectors existed. The slowness of disbursement, together with criticism of Attali's management style and lavish spending, resulted in Attali's resignation in 1993.[58] His replacement was Jacques de Larosière, former managing director of the IMF and chairman of the Bank of France. Under de Larosière, and his successors Horst Köehler and Jean Lemierre, the

Bank's basic mission remained the same, with shifts taking place in terms of the countries of emphasis, strategies for promoting private sector development, and the types of financial tools used to do so. For example, the Bank shifted its focus eastward, to pay more attention to the less reform-minded, more difficult states in the region. It has turned more toward the development of financial sectors, small- and medium-sized enterprises, and the use of financial intermediaries as strategies to promote private sector development. Its tool-kit of financial instruments has expanded to place more emphasis on equity products versus loan finance, and it has increased its involvement in buying stakes in financial institutions in the region.

Birth of Environmental Initiative

The EBRD is the first MDB to be born with an environmental mandate, which was seen by NGOs as a strongly positive move, even if the mandate itself was broad. The goal of promoting "environmentally sound and sustainable development" found its way into the Bank's founding articles as a result of a consensus among major donor countries on the importance of the issue. An NGO campaign was also launched to encourage the new Bank's charter to include an environmental mandate, and its efforts fell on receptive ears.

The United States was particularly forceful in pushing for the inclusion of the sustainable development language in the charter. "With respect to the environment, there is no controversy," wrote U.S. Treasury Secretary Nick Brady during the negotiations on the EBRD. "All of the participants in the negotiations recognize the need to improve the environment in Eastern Europe; and there appears to be a broad consensus among them for strong and effective action. The U.S. negotiators have strongly supported the inclusion [of the language in Article 2(vii)] . . . and will insist that it be included in the final version of [the Bank's] charter."[59] Yet the United States was also aware that the language of the environmental mandate was ambiguous, and argued for the rapid introduction of specific regulations and guidelines as a way to translate the mandate into "strong, effective operational practices."[60]

This environmental policy initiative was aided by the fact that environmental issues were high on the policy agendas of donor agencies as well as recipient countries in the region. The fall of the Iron Curtain revealed the most polluted countries in Europe, and environmental groups in the Eastern Bloc were among the actors pushing to dismantle the communist regimes.[61] Indeed, the United States' aid and assistance initiative for Eastern Europe, the Support for East European Democracy (SEED) Act of 1989, identified environment as one of the key sectors in need of immediate assistance and long-term strategies in the region owing to the extent of environmental degradation throughout CEE and its impact on human health. Also that year, President George Bush announced an East European Environmental Initiative, which included the establishment of the Budapest-based Regional Environmental Center (REC), a nonprofit organization set up to increase environmental awareness in the region, whose charter members included other major Western donors as well as the EC.[62]

In Western Europe, meanwhile, the EC set up its PHARE program of assistance to Eastern Europe in 1989 to coordinate assistance by the G-24 countries (the OECD nations) to East Central European governments in the "high priority" sectors of agriculture, industry, energy, privatization, environmental protection, and financial services.[63] Environment was chosen as an issue because of the EC's concern about the extent of environmental degradation in CEE, the lack of resources available to the region's governments, the EC's desire to create a "level playing field" for trade between its members and CEE, and the transboundary nature of much of the pollution in CEE.[64] Because coordination proved to be too difficult given differing assistance agendas among G-24 states, and U.S. discomfort with the EC Commission as coordinator, PHARE quickly became the EC's own vehicle for assistance to the region. Bilateral European donors also included environmental components in their aid agendas.

Lobbying by NGOs certainly helped promote the environmental issue in the EBRD negotiations, although unlike the case of the World Bank, they did not play a seminal role in putting the issue on the table. The U.S. offices of NGOs such as World Wildlife Fund (WWF), Friends of the Earth (FOE), and the Center for International Environmental Law (CIEL) already had allies within the U.S. Treasury, the EPA, and the State Department who agreed with their aims. In fact, it was a Treasury official who approached his NGO friends with the idea of promoting the concept of sustainable development in the EBRD's founding articles.[65] NGOs from West and East were also able to meet with Attali in November 1990 to discuss their policy goals. Attali, said one NGO activist, was receptive to the idea of environmental considerations being an integral part of Bank policy, but at the same time he was most interested in creating a Bank that was fast-acting and nonbureaucratic. As a result, some of the NGO demands—such as time for public comment on projects—would not be attractive to Attali, since they would slow down the project cycle.[66]

Several other countries on the EBRD's board were also strongly supportive of an environmental mandate. These included Germany, Austria, the Netherlands, and the Scandinavian countries—all countries with relatively strong domestic environmental movements. Another set of countries, including Canada, the U.K., France, and Italy, was more neutral; although they were not indifferent to the Bank's environmental policies and goals, other priorities ranked higher. The countries of operation, in turn, were most cautious and were also generally neutral about addressing these issues, since environmental issues receded as a political priority shortly after the regime changes.

General Environmental Policy Objectives

During its first year, the EBRD developed a set of environmental "policy priorities," which were published in early 1992. In *Environmental Management: The Bank's Policy Approach,* the EBRD outlined its intent to place "environmental issues at the forefront of its efforts to promote sustainable economic growth" and listed six priorities to allow it to play a "leadership role in environmental recovery of the region."[67] These priorities were:

• to help countries develop environmental policy, including assisting development of legislation, appropriate emissions standards, and institutional capacity to monitor and enforce policies;

• to promote market-based and other economic techniques to address causes of environmental degradation;

• to encourage development of environmental goods and service industries and investments in environmental technologies, and to fund environmental infrastructure projects;

• to initiate or support studies and programs that address regional and national environmental problems;

• to adopt "adequate environmental assessment, management planning, audit and monitoring procedures" in all of its activities; and

• to promote the adoption of procedures for "provision of information to, and consultation with, all levels of government and the general public—especially potentially affected parties—concerning environmental matters."[68]

In late 1996, the Bank announced a revised environmental policy, fashioned as a result of debate and discussion among the board directors, the Bank's environmental appraisal unit, and other senior Bank officials, as well as several NGOs. Some shareholders, such as the United States and the Netherlands, were keen to strengthen the Bank's use of specific environmental standards in its work, such as national or EU standards. In the past, the use of standards was more ad hoc, with the appraisal unit deciding whether local standards or some other set of standards would be most appropriate in a project. Some Bank staff argued that using EU standards, for example, was "silly" as well as potentially highly costly for countries in the region that were not prospective members of the EU. The argument was that high standards in some cases might be too difficult or too costly to implement, while lower standards could still result in an outcome that was a vast improvement to the pre-project scenario.

For example, in the case of emissions of air pollutants, there is often a space between existing emissions and future emissions under EU standards. This means that a power plant rehabilitation, resulting in vastly improved energy efficiency and lower emissions at a similar level of energy production, might be too expensive to undertake if higher standards meant the purchase of very expensive equipment, such as flue gas desulphurization (FGD) technology. The pro-higher-standard coalition on the board, in turn, argued that an ad hoc approach to standards was unacceptable. The resulting compromise, five drafts later, was a policy stating the Bank would make sure its loans follow national or EU environmental standards, or where there were no EU standards, follow World Bank standards. The idea was to use EU standards as a common reference point, but keep the flexibility to use other standards where EU standards were nonexistent or did not make sense for the country.

The new policy continued to emphasize the importance of the environmental appraisal process, as well as the EBRD's intention to "realize additional environmental benefits through its operations, particularly if they also provide economic benefits."[69] In particular, it would seek to play a role in financing projects that alleviated "severe environmental problems."[70]

It aimed to describe the environmental implications of its work in the strategies it developed for individual recipient countries and sector work. In this area, the EBRD would draw on work done by other institutions, such as the World Bank's National Environmental Action Programs. It also emphasized the EBRD's appraisal procedures as a way of determining whether a project should be financed or not, and if it is, how environmental issues should be incorporated into its design.

The basic components of the EBRD's environmental policy—emphasizing environmental procedures, funding for specific environmental projects, and some environmental policy work—broadly echo the World Bank's policy. Yet, the scope of the EBRD's actions is narrower, and subsequent chapters will show that the EBRD's private-sector, demand-driven character has provided relatively fewer incentives for its bankers to produce stand-alone environmental projects, with the exception of its municipal environmental unit and its small energy efficiency unit. At the same time, the EBRD has made advances in its environmental activities since its birth, as it has learned how to improve the fit between its policies and mission while revising some key policies and procedures.

EIB

Basic Mission and Lending Policy
The EIB was created to be the EC's long-term lending institution by the 1957 Treaty of Rome, which established the European Community. The six founding EC states were keen on providing financial assistance to the poorer regions, which at the time was focused on Italy's *mezzogiorno* region.[71] The Marshall Plan had ended in 1951, and European governments were still interested in stimulating capital investment. The EIB's job was to "contribute to the balanced and steady development of the Common Market in the interest of the Community."[72] With an initial capital of $1 billion, the Bank would raise money on international capital markets to finance projects in less developed regions of the EC, with the broad goal of undertaking projects "called for by the progressive establishment of the common market," or projects of joint interest to several member states, which could not be financed by means available to individual countries.[73] The Bank would lend money to public sector bodies or large industrial companies to finance projects, although beginning in

the late 1960s, it also began to make loans to financial intermediaries for on-lending to small-scale projects.

Article 18 of the Bank's statute gave the Bank the possibility of lending outside of the EC, in situations where directed to do so by its shareholders.[74] The Bank began its operations in 1958 in Brussels, but subsequently moved to Luxembourg, where it is across the street from the European Court of Justice.

The EIB is owned by its EU member states. Like the other two MDBs, members subscribe to the EIB's capital, and most of the Bank's financial resources are raised on international capital markets, where the EIB is the world's biggest borrower and lender in Euro. Of the three MDBs, the EIB is the only bank that lends most of its money to the countries that are its major shareholders. This means that the distinction between "donor" and "borrower" so visible at the other two MDBs is less distinct here. Indeed, since at least 1990, on average around 90 percent of the Bank's loans were made within EU member states. More than half of the Bank's loans generally go to Greece, Spain, Ireland, Portugal, southern Italy, and eastern Germany.

The Bank's Board of Governors is composed of member state finance ministers, and its Board of Directors represents member states as well as the European Commission. Whereas capital subscription accounts for voting share at the other two MDBs, the link is slightly less direct at the EIB. In this case, the level of country subscription accounts for how many directors a country has, versus a certain percentage of votes, and decisions are taken by a simple majority. Thus the most powerful countries on the EIB's board are Germany, France, Italy, and the U.K., which each subscribe 17.7 percent of the Bank's capital, and each has three directors. Spain has two directors, and the rest of the member states have one each. The European Commission also has a director on the EIB's board, although it does not subscribe to the Bank's capital.[75]

Governance issues are discussed in depth in chapter 4, but it is worth noting here that the EIB is the most autonomous of the three MDBs. It has the odd distinction of being both a legal entity independent of other Community institutions and also an instrument of the Community. It is thus responsible for carrying out Community policy (although chapter 4 explains that how the institution actually does this is often open to interpretation).[76] Where capital replenishment times have been a way for donors to exert influence at the World Bank and EBRD, the EIB has had only eight capital increases in its lifetime.[77] The Bank's unusual character—legally independent but an instrument of the Community, with major lending and borrowing countries comprising the same countries—has shaped its behavior in some important ways.

It has preferred throughout most of its history to act as much like a nonpolitical lending institution as possible, while at the same time it must take on new policy mandates given to it by the Council. The EIB has not had a policy voice like the other two MDBs. This makes sense when one considers that its shareholders would have little incentive to create an institution that tells them what to do. As a result, the EIB has no separate policy research arm or country lending strategies, and it does not directly promote a particular set of its own policy ideas or innovative types of loan conditionality.[78] "Brainwork," said one senior EIB official, "is done elsewhere."[79]

Lacking a policy voice and preferring to be "demand-driven," the EIB has the lowest profile among the three MDBs, and has for most of its life astutely avoided a political role of its own. It prefers not to think of itself as a "development" institution.[80] Historically, the Bank has been quiet and lean, although it began to assert itself more under Sir Brian Unwin, president of the Bank from April 1993 to December 1999. Certainly, its political profile does not match its lending prowess. While the EIB has an EBRD-sized staff of just over 1000, approximately 1/9th the size of the World Bank's staff, the EIB now lends more each year than the World Bank—$34 billion in fiscal 1999 compared with $29 billion for the World Bank, for example. The Bank's capital has increased significantly over the years, doubling from 29 billion to almost 58 billion euro in 1991, and enjoying another large increase from 62 billion to 100 billion euro in 1999.

Although the EIB has not evolved as dramatically as the World Bank has during its lifetime, it has seen a considerable expansion in its geographical scope, lending program, and lending volume. While the bulk of its loans are made to EU member states, the EIB is a global actor, operating in 120 African, Caribbean, Asian, Latin American, Meditterranean, and other states that have signed cooperation or association agreements with the EU.[81] The geographical scope of the EIB's lending reflects its position as a player in the EU's development policy, and the EU's interest in providing assistance to countries that are considered to be of "more direct economic significance" to the Community.[82]

The Bank entered the Central and Eastern Europe arena when its Board of Governors authorized it in 1989 and 1991 to lend 1.7 billion ECU for projects in Bulgaria, Hungary, Poland, Romania, and what is now the Czech Republic and Slovakia. That lending has increased significantly over the years, and the countries covered expanded to include the three Baltic states, Albania, and as of 2000, Bosnia-Herzegovina and FYR Macedonia. Lending for the 1997–2000 period totaled 3.5 billion euro, and was supplemented by an additional 3.5 billion euro pre-accession lending facility for CEE countries and Cyprus to help these countries prepare for integration into the EU.[83] For the period 2000–2006, the Bank is authorized to lend an additional 8.7 billion euro to the ten applicant countries, as well as Albania, Bosnia-Herzegovina, and FYR Macedonia, plus an additional 8.5 billion euro in a new pre-accession facility for candidate countries for the period 2000–2003.

The Bank's lending program has also grown over the years with the addition of new member states, but the scope of projects it funds looks remarkably constant. From its early days onward, within the EU the Bank focused on infrastructure development, particularly the areas of transport, energy supply, telecommunications, and drinking water and waste water treatment. Unlike the World Bank, the addition of new sectors of competence is rare at the EIB. There are only a few prominent examples of this in recent years. At the behest of European heads of state, the EIB was given in 1992 a more public mandate to accelerate lending to trans-European networks (TENs) and infrastructure projects in railways, roads, ports, airports, telecommunications, and energy transmission and distribution, which would establish better and more efficient links between member states. This was essentially asking the EIB to devote more energy to its strong lending areas. More recently, the Bank was asked to take on lending in the health and education sectors, as a result of decisions adopted by the Amsterdam European Council in 1997.

The EIB's lending policies are broadly more flexible than those of its two counterparts, in the sense that it can lend money to either the public or the private sector, with no explicit rule on what share either sector should receive. Like the other two MDBs its statutes stipulate that it is not supposed to compete with the private sector, but EIB staff acknowledge that such competition is difficult to avoid in practice. It can contribute up to 50 percent of a project's cost and can lend to smaller projects (under 25 million euro) through financial intermediaries.

Birth and Evolution of Environmental Initiative

In contrast with the other two MDBs, the EIB has never faced strong pressure to change the ways it takes environmental issues into account. No single shareholder launched a major campaign to push the Bank in new directions. The first attempt by environmental NGOs to launch an EIB campaign occurred in 2000, and it is too early to see the results. The EIB first produced a very modest internal environmental "policy" document in 1984, which broadly echoed the basic policies of the other two MDBs by stressing the importance of considering environmental impacts of projects and financing activities to protect the environment, in areas defined under Treaty Article 130. Some observers, and even Bank officials, would not refer to the document as a "policy" since it reflected the Bank's commitment to follow Community policies rather than developing its own, or showing more explicitly *how* it would translate EC goals into its own internal rules.

The EIB did produce a document on its environmental goals in the absence of any NGO campaign or pressure from Bank shareholders. Yet, the result was a policy document outlining basic goals that were neither a departure from existing Bank practice nor accompanied by any major changes in the institution's structure, procedures, or the way it carried out its basic activities. The new policy was an initiative of Hellmuth Bergmann, the head of the Bank's Technical Advisory Service, then home to the Bank's engineers, who appraise the technical aspects of the Bank's projects.[84] Bergmann, a German agricultural engineer, was responsible for general environmental issues at the Bank and felt the EIB should be more active in addressing these issues.[85] Bergmann was supported by the German and Dutch directors, while other member countries were more or less neutral.[86] The board of directors took up the issue and developed the ideas proposed by the Bank's staff. The board recommended that the Bank "continue to adopt a cautious and pragmatic approach" with respect to its environmental activities, "by, on the one hand, financing economically viable projects furthering the struggle against pollution and, on the other, exercising close vigilance over the projects that it does finance, to monitor their potential effects on the environment."[87]

The policy emphasized what was already implicit in the Bank's work: that it would enforce European Community and national regulations on the environment, as it was supposed to do in all of its work. Where there were no binding regulations, the Bank would consider the "overall impact on the environment when assessing the economic viability of a project" and make "a concerted effort to raise awareness of environmental problems among promoters with the aim of getting them to adopt the least polluting design they can afford without compromising the economic return on the project." This also reflected previous actions, since the Technical Advisory Unit had been assessing the environmental impact of projects since 1972.

The policy also called for the Bank to finance activities that would protect the environment, in areas defined under the broad Treaty Article 130. In addition, the Bank would be able to provide additional financing of up to 10 percent of a project's cost if the project installed antipollution equipment offering greater protection than required under existing standards. For projects outside the Community, the policy said the Bank "should refrain" from financing projects that "seriously transgress internationally accepted standards."[88]

In its basic outline, the EIB's environmental policy document had the same fundamental goals as those of the other two banks: to minimize the environmental impact of all projects and to fund some projects aimed specifically at protecting the environment. However, the EIB's policy goals were the least aggressive of the three. Its emphasis is similar to that of the early World Bank goals—to focus on mitigating potential environmental harm, in this case by following national or European standards. The EIB also often becomes involved in projects in the later stages of their development, which can make it difficult and costly for its technical team to introduce modifications.[89]

Although the policy called for the EIB to fund "environmental projects," this again reflected what the Bank was already doing. Indeed, the EIB has been actively involved in funding the types of infrastructure projects that MDBs have historically undertaken for economic development purposes, which also have environmental benefits, such as water supply, waste water treatment, energy projects that use air pollution control equipment (such as FGD equipment for coal or lignite-fired plants), waste management and urban transport (such as metro or tram expansion, and suburban rail service improvement). Ultimately, the EIB's environmental policy document was not a forward-looking statement aimed at moving beyond standard operating procedures and habits common among MDBs.

The strengthening of European environmental policies in the 1980s and 1990s has affected the Bank mainly in giving it more regulations and directives to follow, but to date it has not appeared to have a significant impact in *how* it goes about its business, which is discussed in chapter 4.[90] Environmental concerns were given a big boost in Europe in 1986, when the Single European Act (SEA) incorporated environmental policy into its amendment of the Treaty of Rome, thus making environmental policy a formal competence of the Community.[91] The 1992 Maastricht Treaty, in turn, made explicit the task of promoting sustainable growth respecting the environment in its Article 2 and called for environmental protection requirements to be integrated into other Community policies.

General Environmental Policy Objectives
The EIB produced a revised environmental policy document in 1996, which was again minimalist. It reiterated that funding projects "where the main objective of the investment is environmental" is "a major activity of the Bank."[92] It also noted the importance of considering environmental issues in its appraisal of all projects, with a goal of conforming to European Community and/or national legislation. It emphasized, like the old policy document, that projects must rely, at minimum, on EU environmental standards, while it may take a more flexible approach in its work outside the EU, with a goal of ensuring "good environmental practice at affordable costs." EU standards guide the Bank's work within countries aiming to join the EU. Most of the remainder of the policy document describes the Bank's project and appraisal procedures, which are analyzed in the subsequent chapter. Compared with the other two MDBs' policy goals, this Bank has no ambitions to be involved in policy development issues or to broadly assess a country's environmental situation in analyzing its lending portfolio.

Conclusion

This chapter focuses on the sources of an MDB's policy commitment to environmental issues and highlights the ways in which pressure from major shareholders is a determining factor in the depth of this commitment.

Pressure from the United States was particularly important in explaining the stronger environmental commitments of the World Bank and the EBRD, relative to the EIB. The existence of a set of environmental problems attracting the attention of NGOs and policymakers provided the window of opportunity for pressure to be translated into bank goals. These findings have important policy implications for actors such as NGOs that are seeking to influence MDB environmental behavior. The main lesson is that pressure is best focused on shareholder governments versus bank staff, while leverage is increased during times of capital replenishment and when there is a set of environmental problems or issues that can capture policy attention. NGOs may build up contacts and share ideas with bank officials, but this is not what will drive fundamental changes in bank policy.

The EIB's weak environmental policy goals can be explained by the absence of any environmental champions on its board and the lack of any environmental disaster or particular issues that might be publicized by NGOs. A weak environmental mandate immediately lowers expectations of the EIB's environmental performance. We would expect to find few incentives for staff to address environmental issues beyond a "light green" criteria, and a project portfolio that reflects traditional MDB emphasis on infrastructure projects. In other words, in the three-stage analytical framework discussed in chapter 1, the EIB effectively is mired at the policy objective stage. "Environment" projects at the EIB will be largely old wine poured into new bottles.

The findings of this chapter also disprove hypotheses that might focus on the age or size of an MDB as determining factors that shape its environmental commitment. The youngest of the three MDBs has a weaker commitment than the oldest, but a stronger one than its other older colleague. The EIB and World Bank lend similar amounts each year, while the EIB and EBRD have a similar number of employees. These factors clearly do not correlate with the strength of an MDB's decision to address environmental issues in its work.

Turning now to the next stage of the policy process, we will see how institutional variables play a causal role in determining how these broader policy goals are translated into institutional behavior.

4

Policy Process: Institutionalizing Environmental Objectives

This chapter examines why external pressure for change is a necessary but not sufficient factor in explaining how the MDBs respond to their delegated authority. All three banks have diffuse governance structures, which privilege the role institutional variables play in determining the capacity of an MDB to pursue its environmental goals. Shareholders set the basic rules and design of the institutions, are influential at the strategic level (such as agreeing to new mandates), and ultimately must approve all loans, but their control is limited by the banks' relative autonomy and the fact that new mandates do not always fit well with existing institutional design and incentives. Donor preferences on new mandates, as a result, are not sufficient in determining which activities and projects a bank is willing and able to offer recipient countries. In the language of principal-agent theory, there is considerable "agency slack" in all three MDBs.

The institutional variable that most influences MDB responses to policy goals, given weak governance structures, is how demand-driven or bank-like the MDB is designed to be. As noted in chapter 1, all MDBs are designed to function in part as financial institutions and in part as development agencies that promote a variety of mandates to shape policy reform in recipient countries. However, they differ in the degree to which they emphasize their banking goals or their nonbanking goals. An MDB designed to operate more like a financial institution than a development agency will be more driven by client demands than broader policy interests. In what Naim calls the "Bank-as-bank" model, the dominant goal of an MDB would be to focus most on its long-term financial integrity and fiduciary responsibility to its shareholders.[1] This is, however, hypothetical, since no MDB is without some "development" goals. Further, MDBs tend to finance projects that private sector banks may find too risky, while at the same time they typically impose more conditionality in their loans than a private sector bank. The issue is which goals the MDB emphasizes and how it juggles activities where goals conflict or do not complement each other. The more an MDB conforms to the "Bank-as-bank," model, the less important is the objective of promoting far-reaching policy changes in the recipient country. The result is fewer incentives for such an MDB's staff to actively identify and design projects with primary environmental goals or even significant environmental components unless these are specifically sought by the recipient.

By contrast, a less banklike MDB will likely contain more incentives for staff to link loans to broader policy goals, such as environmental protection and reform. However, this chapter also argues that the less bank-like and demand-driven an MDB is, the more it must struggle to harmonize conflicting goals and incentives. Environmental policy objectives and activities do not always fit well with other economic development or banking goals, as noted in chapter 1. Conflicting incentives, in turn, may arise when project managers, seeking to bring creditworthy, bankable projects to their boards of directors in a timely fashion, balk at the requirements of environmental assessment procedures that can add delays to projects or require costly changes in their design. Therefore, even where the right institutional variables exist to promote greater awareness of environmental issues, they do not always work as envisioned. While shareholders determine the key parameters of institutional design, there are often gaps between how they have constructed the institution and what they ask it to do. As a result, the institutional "hardwiring" of mandates must not just be described and behavior then inferred; it must be analyzed in order to understand how it shapes behavior.[2] Although it is commonly accepted in the environmental policy community that MDB (particularly World Bank) environmental action cannot be understood by examining the institution's stated rules, it is less clear how exactly institutional factors shape environmental behavior.

Variation in the types of activities the World Bank, EBRD, and EIB are likely to pursue can also best be explained by examining how each MDB responds to its dual personality as financial institution and development agency. This chapter shows the extent to which the World Bank most emphasizes its role as a development agency but struggles more in its attempts to harmonize multiple mandates. The EIB is most banklike, thereby sticking to a narrower range of more traditional MDB lending activities and facing fewer conflicting incentives in its work. The EBRD is between the two, since it was set up to focus mainly on private sector development, but its design and incentive systems do include components that encourage its staff to address particular policy issues in greater depth than a private bank would do. In the case of the EIB, because of a weak donor commitment to environmental issues, we would not expect the institution's design, rules, or procedures to be shaped in ways that encourage the identification and design of projects with primary environmental goals. We would expect to see more institutional incentives to do so in the World Bank and EBRD.

This chapter begins by comparing the governance structures of the three banks, highlighting the banks' relative autonomy from the governments that "own" them. In social scientific terms, it shows how agency slack is a constant variable among the three banks. The subsequent three sections on each MDB examine the evidence for how banklike each MDB is and how this influences staff incentives on environmental issues. Five major indicators reveal the degree to which an MDB emphasizes one side or another of its dual character. These are: institutional mission, structure, financial tools, procedures, and porousness. *Mission* refers to the institution's overall aims. This determines not only the major issues an MDB seeks to address, but also the types of actors (state or private sector) to whom it provides financial resources and advice. The more demand-driven an MDB is, the more it is responsive to client interests, and the less it seeks to impose its own policy prescriptions or reforms in its loans or other activities. *Structure* refers to operational design factors and staff composition and placement. An important component of staff composition is the existence and location of "green bankers," or bank staff whose job is to identify or appraise environmental projects and to make more of a point than their colleagues to pursue environmental goals in their lending or policy work. The "green bankers" are the people who have particular incentives or training to give environmental issues more prominence in their work, and for whom concern about environmental issues is embedded in their thinking.[3] The composition and location of these "green bankers" can profoundly influence how an MDB's portfolio of project and other activities reflects its environmental goals. As expected, less banklike MDBs have more "green bankers."

The *financial tools* determine what types of projects the bank can fund and also who the recipients are; this variable is not well explored in the literature on international assistance.[4] *Procedures* specify how the bank intends to pursue its various goals. Procedures most relevant to the bank's environmental behavior can be found within overall project cycle processes and specific environmental due diligence requirements.[5] Finally, *porousness* is an important but less tangible factor. It refers to an MDB's openness to outside scrutiny and pressure, which is partly embodied in its procedures but may extend beyond them. This openness is an indicator of MDB transparency and can influence the ways an MDB considers environmental issues.[6] MDBs that behave most like private sector banks will be more attached to rules of confidentiality and less open to interacting with or involving nonrecipient actors.

It is helpful to separate these variables analytically, but in practice they are clearly entangled and interact in complex ways. For example, an MDB's mission shapes its policies and procedures, as well as the way it is structured and the types of financial tools it uses. Staff composition, too, shapes procedures (e.g., environmental appraisal procedures are likely to be more developed where there are environmental staff to design and implement them). Further, the existence of "green bankers" can also be traced back to shareholder pressure. Feedback loops undoubtedly exist in various areas as well, such as instances where MDB research influences donor priorities, which in turn can shape staff composition or procedures. However, the fact that these variables are to some degree entwined does not detract from the importance of analytical attention to their impact on how demand-driven the institution is, and how this affects its environmental behavior. The chapter concludes by hypothesizing how variation in the ways the MDBs juggle their dual personalities may affect their activities in Central and Eastern Europe.

Diffuse Governance

All three MDBs have similar basic governance structures—a three-tiered structure consisting of a board of governors, a board of directors, and some type of management operations committee. The board of governors, generally composed of finance ministers or other ministerial level officials from all of the bank's member states, meets once a year.[7] The governors of countries belonging to all three MDBs tend to be the same officials. Their job is to lay down the overall directives for the MDBs, including decisions on capital increases, approving the annual report and other financial statements, new membership, and so on. The presidents of the banks, in turn, are the banks' chiefs of staff, and they participate in the meetings of the boards of governors and chair the board of directors meetings.

Shareholder countries are represented at the banks through their boards of directors, to which the governors delegate much of their power. The board of directors is the body responsible for an MDB's general operations and for approving all loans and major bank policies. It is the primary channel through which the shareholder nation-state principals are involved in the MDB's activities. In effect, the "principals," who run the MDBs—that is, their boards of governors and directors—are themselves a diffuse set of "agents" representing their home governments. Finally, each bank has some type of management committee that is responsible for its day-to-day business. These committees are the main gatekeepers that decide which loans and programs go to the board for approval.[8] The following section describes important characteristics of the board of directors' composition, representation, and activities and shows how these attributes result in a weak set of collective principals with diffuse preferences for all three MDBs.

Composition and Representation[9]

The World Bank's board consists of twenty-four executive directors (EDs) representing 180 countries, and the EBRD's contains twenty-three, representing fifty-six countries and two international institutions.[10] In each bank, only a handful of the directors represents one country. At the World Bank, the United States, France, Germany, Japan, and the U.K. have their own directors, as do Saudi Arabia, China, and Russia. In the case of the EBRD, the United States, France, Germany, Japan, the U.K., Italy, and the Netherlands have their own EDs, as do the EIB and European Commission. Beyond this group, all the other countries are grouped into constituencies. At both banks, recipients are often grouped together, such as the EBRD's ED office for Hungary/Czech Republic/Slovak Republic/Croatia. However, the ED constituency offices representing Central and Eastern European countries at the World Bank are a mix, with a donor country providing the director.[11] A few constituencies at the World Bank consist of over twenty countries. Sometimes members of the constituency have opposing interests and it is up to the ED to determine how to vote. In both banks, voting power is related to the share of a country's contribution to the MDB's capital.

Given these characteristics, most of the countries belonging to these two MDBs do not have a direct voice or vote. Many are not really represented at all at times, since the staff of a multiconstituency ED office is not large enough to accommodate a representative from each country of its constituency.[12] Major donors therefore have a louder voice at the World Bank and EBRD not just because they have a greater voting share, but also because they have single-constituency offices.[13]

The case of the EIB is different in the sense that it has no multi-constituency offices. The larger European Union (EU) countries have more directors than other EU countries, and each director has one vote. Germany, France, Italy, and the U.K. each have three directors, while Spain has two, and the remaining countries have one each. The European Commission also appoints its own director to the EIB. The size of the EIB's board currently stands at twenty-five, with thirteen alternates, and will likely grow even larger as a result of EU enlargement. This composition does not necessarily give member states more control over running the Bank; the EIB's relative autonomy is augmented by the fact that its board is non-resident. While the resident boards of the other two banks meet twice a week (World Bank) or twice a month (EBRD), the EIB's board travels to Luxembourg to meet around ten times a year.[14]

Who is actually represented by the board? In the case of the World Bank and EBRD, the diffusion of shareholder preferences is enhanced by the fact that the EDs themselves are agents, representing a variety of home government ministries. Even for ED offices representing only one country there are varying degrees of contact with home ministries, with many countries giving the ED office a great deal of discretion in voting decisions.

In the case of the G-7, for example, the United States is widely acknowledged to have the most proactive home ministry (Treasury), which gives U.S. EDs relatively more precise instructions in what to push for and how to vote on the boards of MDBs where the U.S. is a member.[15] While the EDs at the World Bank and EBRD are presidential appointees, their alternate directors are from the Treasury, and the ED offices receive guidance from the Treasury, through its MDB office with a staff of around twenty. The U.S. Governor of both MDBs is the Treasury Secretary.

The U.S. ED offices have instructions on how to address priority issues at the board level, and even mandatory voting positions on a number of issues. Instructions are for issues such as promoting energy conservation and end use efficiency, or for opposing loans that would support the export of surplus commodities that would hurt U.S. producers.[16] Legislatively mandated voting includes issues such as human rights (the United States must vote "no" or abstain from voting on loans to countries that violate human rights unless the assistance is for basic human needs); or terrorism (the United States must vote "no" on loans to countries that the Secretary of State determines support international terrorism).[17] For environmental issues, there is an environmental review unit within the Treasury's MDB office whose job is to assess environmental implications of proposed projects. Much of the advice from Treasury is in the form of instructions on individual projects. Treasury and ED office officials note, however, that the process is often consultative, and that ED offices have input on the decisions. Other United States government agencies may also provide their input on specific projects or policies, although where there is disagreement, the Treasury has the final say.[18]

EDs of other G-7 countries report to their finance or treasury ministries or to their development ministries. Germany's ED at the World Bank, for example, represents its Ministry of Development Cooperation (BMZ), while its alternate comes from the Ministry of Finance. At the EBRD, the German ED is from the Ministry of Finance, while his alternates have been from both the Ministry of Foreign Affairs and the Ministry of Economic Affairs. The British ED at the World Bank is from the Treasury, while the alternate represents Britain's aid agency, the Department for International Development (DfID).[19] At the EBRD, the converse is true; Britain's ED is from DfID, and the alternate comes from Treasury. Occasionally other ministries are also represented. Canada, for example, has had a World Bank ED from its environment ministry.

Members of G-7 ED offices report that it is not uncommon for different agencies representing one shareholder country in the ED offices to have disagreements among themselves over policy directions. This is unsurprising given the different portfolios of finance and development ministries, but it further illustrates how diffuse shareholder instruction or preferences can be.

Consultation with home ministries varies among countries, and also with respect to the types of decisions being made. For some countries, such as Italy and France, specific voting instructions are rare at the World Bank and generally arise with respect to controversial projects. German and Canadian ED offices, in turn, receive more guidance from home ministries.[20] Interviews with members of G-7 ED offices at the MDBs and officials in their home ministries highlight the fact that the relationship between the two is often flexible and consultative. Often, the ED offices make their own decisions. British ED office officials, for example, note that guidance on project votes is less common than guidance on policy discussions. For projects, the World Bank ED office will often make decisions on its own, unless it opposes some aspect of a project. In that case, it will try to clear its opposition in advance with the DfID. In the case of policy decisions, the ED office will receive instructions from London, although it still may take the lead in developing the U.K. position, with clearance from London.[21] U.K. and other ED office officials point out that a two-week turnaround time from when they receive project documents to the day of a board vote is often insufficient for their home ministries to produce thorough briefings.

An official in the German ED office at the EBRD, in turn, stressed that while strict instructions come from the Ministry of Finance, the ED office has a great deal of flexibility in its actions. There are no mandated votes for Germany and most other countries, unlike the United States. All officials interviewed stressed the importance of personal contacts within ministries.

The executive directors to the EBRD from the European Commission and the EIB have even greater discretion in how to vote. The Commission officials, for example, are representatives of the Commission's Directorate General (DG) II, but the EDs do not ask for instruction on many issues. In cases where instruction is sought, it comes from relevant DGs in the Commission. In some instances, different DGs disagree on what instruction should be offered.

The fact that different EDs seek different levels of guidance from home ministries is not necessarily indicative of agency slack. The EDs may be doing their job as agents, interpreting the wishes of the home ministries without needing direct advice. It is also true that as an ED office gains expertise over time, its director and alternate director know how they are supposed to vote without seeking confirmation. Yet the relative flexibility of many of these offices does highlight a situation of diffuse oversight by member states, where it is unclear how responsive ED offices are to their home governments, or vice versa. "Politics on the board," said an official in an EBRD ED office, "seem more based on personalities than nationalities." ED staff commonly note that the "power of personality" is one reason a country's power to influence the board does not always equal its official voting share.

The case of the EIB differs from the other two banks in ways that increase its relative autonomy. On one hand, some members of the board *are* high-ranking treasury or central bank officials in their own country, versus the more "middleman" position of EDs at the other two banks, who are lower ranking officials representing their home ministries. In recent years, for example, the EIB's Irish directors have been the Governor of the Central Bank and the Second Secretary of the Ministry of Finance's Finance Division. In many cases, the EIB board directors are literally the bosses of the board directors from the same countries at the other two banks. Interestingly, for countries with more than one director on the EIB board, representatives may also come from outside government ministries, and even include private sector actors. The 1995 board, for example, included an economics professor from a Greek university, the head of the board of Germany's public sector bank, Kreditanstalt für Wiederaufbau, and an official from a London-based merchant bank, Barclays de Zoete Wedd.[22] Given that the EIB's board is nonresident and contains representatives for whom EIB board membership is clearly not their only responsibility, it is no surprise that it takes a relatively more hands-off approach relative to the boards of the other two MDBs.

Another aspect of board composition that is a potential source of board weakness is the relatively short tenure of board directors. At the World Bank, directors have two-year, renewable appointments. While some directors stay for many years, Naim has estimated that approximately 65 percent of board members leave after serving less than three years.[23] There is a similar degree of turnover at the other two banks. Board appointments are for five years at the EIB, and three years at the EBRD. While board tenures are also renewable at these two banks, high turnover is still evident. Turnover at both the EIB and the EBRD, for example, has averaged around one-third of the board each year between 1994 and 1996.[24]

One result of this situation of diffuse governance is that the MDB presidents are often powerful actors in their own right, playing a significant role in directing the activities of the banks.[25] Robert McNamara, for example, was clearly the driving force in the 1970s in expanding the World Bank's lending and defining its mission to put poverty alleviation as its highest priority, despite opposition from senior United States treasury officials that the Bank was growing too quickly and not responding to donor criticism.[26] More recently, while James Wolfensohn has been pushed by donors to make the World Bank leaner and more competitive, it is his vision of the Bank's evolving role in the international economy that has determined much of the content of the reorganization. The same power can be attributed to some of the presidents of the other two banks, although there is also evidence that if a president strays too far from donor interests, he is not indispensable. This was the case with the EBRD's founding president, Jacques Attali, who was a major intellectual force behind the creation of the Bank but was forced to resign amid donor displeasure over his management style.

Activity

In addition to issues of composition and representation, other factors contribute to generally diffuse governance structures among the three MDBs. A particularly important one is the way in which the boards conduct their activities. This creates space for greater autonomy of the MDBs' management and staff, particularly with respect to the boards' ability to influence project development. In terms of loan approval, one of the most important activities of an MDB's board, most board members learn about most loans at the end of the project cycle, when loans are presented to boards for approval. In the case of the EBRD and World Bank, this is usually less true for directors representing recipient countries, since they have greater links to and more communication with bank staff involved in developing projects in their countries.[27] Yet, board members from richer countries do have a certain incentive to find out what they can about relevant projects, because these projects are the source of billions of dollars a year in contracts for the private businesses that provide the equipment, construction, or advice to individual projects. Naturally, directors want to help encourage as much procurement as possible to flow to their nation's companies.

Nevertheless, many board members will not have detailed knowledge of a loan until they receive the board documents two weeks before the board meeting. At this stage, after a project may have been developing for well over a year, it is typically difficult for any changes to be made. It is also not always clear how in depth the board's information is. The boards receive shortened, summary versions of the project documents. NGOs often argue that it is easy for bank management to leave controversial issues out of the reports. At the World Bank, it is only since the mid-1990s that board members have even had access to the Country Assistance Strategy (CAS), the main strategy document between the Bank and a borrower where the Bank's assistance levels and goals are laid out.[28] While new information disclosure policies at the Bank have improved this situation since 1994, EDs and their staff sometimes still bridle at their sense that management can call the shots when it so desires. At the same time, board members at all three MDBs sometimes hear about problem projects—or even good projects—in advance. The more active board members will also communicate directly with project managers.

Board voting procedures at all three MDBs encourage an atmosphere of consensus. At all three banks directors rarely have formal votes. Virtually the only time a director will "vote" is if he or she wants to go on the record as saying "no." It is rare for board meetings to have "no" votes, and even rarer for sufficient board opposition to prevent a project from being approved. Management can also reduce the odds of disagreement and conflict at board meetings by withdrawing controversial loans before they arrive at the board. In addition, no one country has sufficient voting power to block a project without support from other shareholders.

At the World Bank, procedures have been streamlined so that some projects no longer even come up for a formal vote. Projects worth a discussion include those occurring in a new World Bank member country, or projects with innovative or problematic components. Where no ED has any matters to discuss, projects are automatically approved. Given the heavy volume of projects going through the World Bank, ED officials say that there is no way they can carefully go through the details of each project. The World Bank, for example, approved an average of 281 projects a year during its fiscal 1994–1999. This compared with around 121 at the EBRD and 365 at the EIB.[29]

Ultimately, MDB executive directors have more power in setting policies and changing institutional design, but less control over the types of projects ultimately identified by bank staff, and the ways in which these projects are designed and implemented. Board members are influential on an ad hoc basis, and generally have the most influence on MDB policy at times of capital replenishment, which has occurred rarely at the EIB and once to date at the EBRD. Leading shareholder countries have had the most opportunities to influence policy changes at the World Bank, where its IDA arm must be replenished every three years. The United States has been the most active among major shareholders in pushing environmental ideas and other policy changes into the World Bank and EBRD, and it is often supported in these efforts by countries such as Austria, Germany, the Netherlands, and Sweden. Several other major countries—including Japan, France, and Italy—are often more passive or neutral on environmental issues. However, the alliances of board members supportive, critical, or neutral to new policy overtures depend on the particular issue at hand.

Every large organization, both public sector and private sector, faces the challenge of balancing the power of its management versus its board. This is the essence of the scholarship on principal-agent relationships. While the MDBs have avoided the extreme poles of principal-agent relationships—that is, where the president and management are completely autonomous from the shareholder-owners, or shareholder-owners micromanage every decision of the institution without utilizing the inherent advantages of agency delegation—the MDB governance structure does allow for a great deal of institutional autonomy.

The next sections of this chapter turn to what happens to environmental policy goals that are brought into the banks via their boards by analyzing the major indicators determining how banklike each MDB is and the impact this "hardwiring" has on staff incentives.

Institutionalizing Environmental Policy Goals

World Bank

Of the three MDBs, the World Bank has the largest number of environmental staff, the most comprehensive set of policies and guidelines, the greatest emphasis on evaluating project implementation, and the most openness to building relations with outside interest groups, such as NGOs. Yet it also faces the most external scrutiny and criticism of the three MDBs, a sign that its intentions have not been well translated into actions.

Mission   As noted in chapter 3, the primary mission of the World Bank has evolved considerably over the years, largely in ways that have moved it away from the "Bank-as-bank" model. Of the three MDBs, the World Bank's mission is least demand-driven, since it seeks to fund projects that meet various programmatic criteria with social, developmental, environmental, or other goals that move beyond the traditional types of bankable MDB projects. It tends to have the most sophisticated types of conditionality attached to its loans, with more examples of less traditional forms of (noneconomic) conditionality and other types of policy influence.[30] This character was underscored by Wolfensohn's 1996 call for the Bank to become a "Knowledge Bank," playing a greater role in disseminating lessons of development, and three years later, the launch of a "comrehensive development framework," a simultaneously ambitious and ambiguous strategy that seeks to encourage a recipient's responsibility for its development strategy in a way that involves not only the government and donor community, but also civil society and the private sector.[31] The Bank's broader mission provides a foundation for the institution to have greater leeway to promote projects and policies that go beyond traditional infrastructure activities. An environmental mandate in an institution with such a mission would appear to have a greater chance to take root than in an institution driven more by traditional banking goals.

The Bank's broad landscape of goals has both positive and negative impacts on the incentives of its staff to carry out environmental objectives. On one hand, the Bank's goal of promoting environmentally sustainable development has resulted in a considerable increase in the resources devoted to pursuing it, including the number of staff working on environmental issues in operations and in research, as well as the number of procedures and policies devised to ensure environmental issues are taken into account in project identification and development. On the other hand, there is concern among analysts of the World Bank that the sheer number of institutional goals creates a situation of mission ambiguity, where multiple strategies can come into conflict with each other and staff have the ability to avoid decisiveness on the issues where strategy conflicts with organizational incentives.[32]

An important recent example of this problem is Wolfensohn's efforts to make the Bank more demand-driven by increasing client-orientation and strengthening the role played by the IFC and the Multilateral Investment Guarantee Agency (MIGA), which insures private sector investors against political risks in developing countries. The goal of client orientation is to better customize country assistance strategies and to build more local ownership of World Bank projects and programs. The goal of strengthening the private sector components is to allow the Bank group to better compete with private sector actors, given the large increase in private sector investment in development countries.

These goals have the potential to clash with the Bank's environmental goals. First, while client responsiveness appears to be a sensible goal, it will reduce the Bank's leverage to push for environmental projects, or to build environmental considerations into other projects in countries where such issues are not policy priorities. Generally, an emphasis on client responsiveness makes it more difficult for staff to build conditionality (political, economic, environmental, or other) into loans. Second, the IFC and MIGA are widely perceived as doing a poorer job than the IBRD in building environmental features into their work. The prospect of the Bank emphasizing guarantees, as part of its strategy of strengthening private sector activities, also reduces its ability to take the initiative on project identification. These areas of potential conflict can be seen more clearly in the following sections.

Structure   Probably the most important characteristic to note about the Bank's structure is that it is far from static and has undergone numerous reorganizations over the years. There have been three major reorganizations alone in the period since 1987, when the Bank began focusing greater attention on its environmental performance. The first of these occurred in 1987, when Barber Conable split staff between those who undertook operational (lending) activities and those who undertook policy research and planning activities. New "Country Departments" were set up within each operational region, combining the separate projects and program departments that existed previously. Operations was put under its own senior vice presidency and covered five regions, which were increased to six when Europe and Central Asia (ECA) was added in 1991.[33] In 1993–1994, under the presidency of Louis Preston, three central, thematic vice presidencies were created to strengthen the Bank's expertise in human resources and operational policy, environmentally sustainable development, and finance and private sector development. Facing pressure from major shareholders to reduce staff and budget, Preston also merged a number of divisions and reduced the number of division chiefs.

Wolfensohn, who took over the presidency of the World Bank in 1995, launched in mid-1997 another extensive restructuring of the Bank, dubbed the "Strategic Compact," as part of his efforts, discussed above, to make the Bank more client-oriented and less bureaucratic. The two important structural changes resulting from this restructuring were the reorganization of the Bank's operational divisions, and a move to decentralize the Bank's Washington-heavy presence by placing more country staff into the field.

In the old Bank, whose basic form was in place since Conable's 1987 restructuring, country departments could contain over 100 people. These departments were broken down into sectoral divisions, where task managers focused on developing projects in areas such as agriculture, environment, trade, infrastructure, and human resources. Task managers were the main Bank officials responsible for shepherding projects through the project cycle.[34] Sectoral experts could also be found in the four regional technical departments, whose job was to work on specific technical issues associated with projects. Under Conable's 1987 restructuring, regional environmental divisions (REDs) were placed in these technical departments to help operations staff design environmental components of projects, as well as to review projects' adherence to the Bank's environmental policies. In having a review role, these RED division chiefs had the power to stop projects from moving forward in the project cycle if they had any concerns.

The major effect of the changes under Wolfensohn was to create vastly streamlined country management units who staff individual projects out of the broader pool of Bank staff, or from the outside. The result was a shift in power to country directors, who control the country budgets. These units contain a handful of people and are organized into six geographical regions: Africa, East Asia and the Pacific, Europe and Central

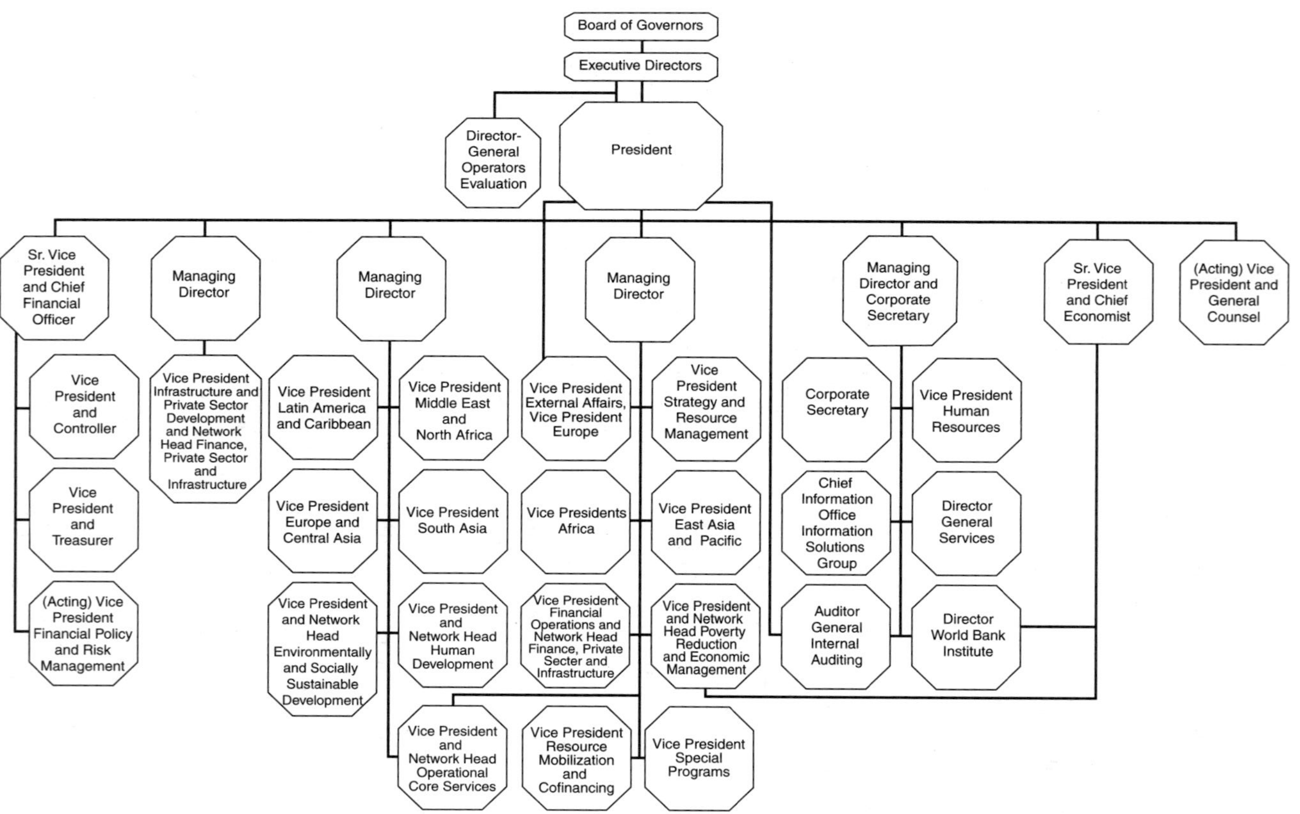

Figure 4.1
World Bank organizational chart, 1991.

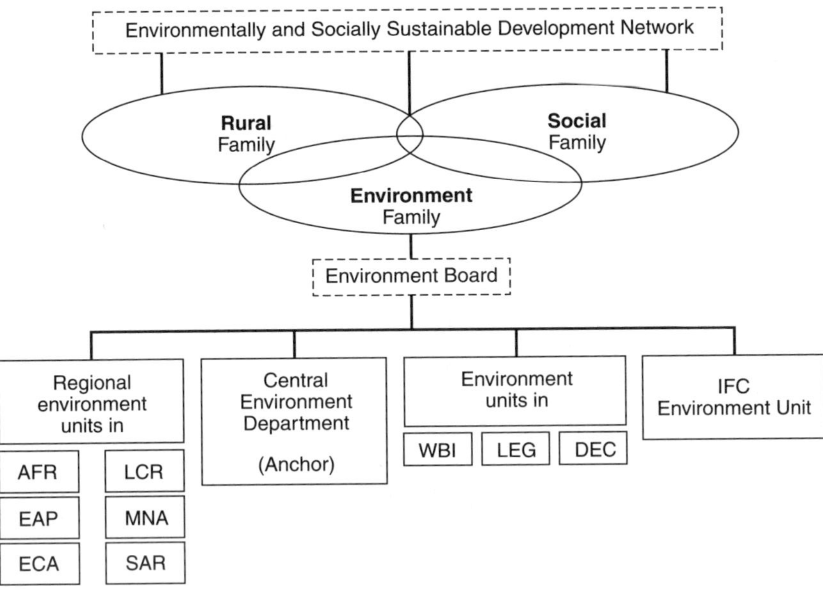

Figure 4.2
Environmentally and socially sustainable development network. *Note:* AFR—sub-Saharan Africa; EAP—East Asia and Pacific; ECA—Europe and Central Asia; DEC—development economics; IFC—International Finance Corporation; LCR—Latin America and the Caribbean region; LEG—legal; MNA—Middle East and North Africa; SAR—South Asia region; WBI—World Bank Institute. *Source:* World Bank, *Making Sustainable Commitments: An Environmental Strategy for the World Bank,* 2001.

Asia, Middle East and North Africa, South Asia, and Latin America and the Caribbean. The small country teams do not undertake projects themselves, but rather advise the country director on broader strategies for the country. The rest of the people from the old country and technical departments are divided into sector groups—one of which is the environment—for each region. These sector groups have larger staffs but small budgets. Country directors hire staff for projects from their regional sector groups, but they also may look elsewhere inside or outside the Bank for the necessary expertise. The reorganization cut the layer of bureaucracy previously staffed by division chiefs, left many managers without jobs, added a new layer of bureaucracy at the top (vice president and managing director level), and also was widely unpopular with staff, many of whom felt it resulted in heavier workloads.[35] It added more layers of complexity in an effort to make it easier for staff in similar areas of expertise to share knowledge across the Bank. The sector groups were made part of four "networks" that link technical staff in the same fields working in different geographical regions on human development, environmentally and socially sustainable development (ESSD), poverty reduction, and private sector development. The networks, in turn, contain "families" within specific issue areas. For example, ESSD contains three "families"—environment, rural, and social—and the environment family includes among its members the regional environmental departments, the central environment department and staff from a few other areas of the Bank.[36]

Location of Environmental Staff
Of the three MDBs, the World Bank has the greatest number of staff whose job is specifically to address environmental issues. For much of the period examined by this book, there were three primary locations for these "green bankers" in the Bank: within the central environmental department, where work emphasized research, policy development, and operations support; within the regional technical departments; and within the country departments themselves, where there were individual task managers responsible for identifying projects with primary or significant environmental components.

Under the most recent organizational changes, the environmental sectoral divisions combined the task managers formerly in the country departments, as well as staff from the regional technical units, and are linked together as part of the environment "family" within ESSD. However, some other environmentally oriented projects are undertaken by staff in other areas of the Bank. For example, water sector projects may be designed by staf in the infrastructure area, which also addresses the urban and transport sectors. At least through the late 1990s, there were around thirty people who worked on environmental projects in Central and Eastern Europe (within Europe and Central Asia region, or ECA), in some capacity or another, plus an additional handful in the central environmental department. One result of these changes was to vastly reduce the importance and power of the central environmental department, which turned its attention toward global and strategic issues.

Impact of Structure and Staffing on Green Activities

The structure of the Bank has characteristics that have both encouraged and inhibited its ability to address environmental issues in its work. The most important factor encouraging the Bank to be active in identifying environmental projects has been the existence of "green bankers"—the task managers looking for such projects.

Generally, Bank officials have strong incentives to get projects to the board in a timely fashion, which has led many critics of the Bank to argue that project quality can be pushed aside. This has been labeled as the Bank's "approval culture," reflecting the fact that task managers have tended to be judged more on project preparation and appraisal versus implementation.[37] In addition, because a project can take years from start to finish, quite often those in charge of the project will have changed jobs before a project is complete. The incentives that create pressure to lend are often not explicit, and management in recent years has sought to emphasize the importance of project quality over lending targets. Still, in practice, Bank staff will be more attracted to relatively large projects that are relatively easier to undertake, with as few complex issues as possible to slow them down. Projects with lower transaction costs are more attractive, which means staff are generally not in favor of expensive, in-depth environmental assessments that may require significant changes in project design. Yet, Bank staff with the specific goal to seek out projects that address a country's priority environmental problems clearly have more incentive than others to do so. This has resulted in an overall World Bank portfolio that contains many projects not traditional even to other development banks, such as environmental management capacity building, natural resources management programs, coastal pollution control, and community forestry, to name some.

Bank staff also design projects under the auspices of the Global Environmental Facility (GEF), which provides grants to developing countries to address four global environmental problems: ozone depletion, diodiversity loss, climate change, and the degradation of international waters. The GEF portfolio at the Bank thus contains projects that go well beyond typical MDB infrastructure projects and cover topics such as wetlands management, biodiversity conservation, and the phaseout of ozone-depleting substances.

Another major structural characteristic shaping the Bank's environmental activities is its work on the "knowledge front" in research, and in training and policy work on the environment for recipients. Staff involved in environmental research were found for many years within the central environmental department, but also within other areas of the Bank, such as the environmental research group under the development economics vice presidency. As noted in chapter 3, this research has included sectoral policies as well as "best practice" guides for Bank work on the management of different types of natural resources and sources of pollution. It also includes the Bank's work on national and regional environmental programs for recipient countries, the development of least-cost, "green" accounting, and environmental impact assessment methodologies, as well as research on global issues such as climate change and ozone depletion. Generally, environmental specialists have penetrated the Bank to a greater degree than the other two MDBs, and their work is no long considered marginal. Yet at the same time, the various structural iterations contain negative features that reduce the ability of the Bank to meet its goals.

First, the links between operations and research have historically been murky and not well integrated. Critics of the Bank have argued that the research and policy arms of the Bank are poorly connected to operations. Rich, for example, argued that the 1987 reorganization created a central department that "inhabited a world of paper . . . while the lending juggernaut lumbered ahead on a separate planet called Operations," and the regional environmental staff were "all but powerless to stop ambitious country directors from riding roughshod over Bank policies."[38] In fact, this assessment is overly pessimistic, since central environmental staff ended up spending more than half of their time supporting country departments, and regional environmental staff were able to stop projects from moving forward, on environmental grounds.[39] Yet, there is much evidence that the policy work and research undertaken by the Bank are often poorly translated into actions by operations.[40] The question is how to assess the importance of the research if often it exists in a vacuum from project work. Critics have labeled the Bank's policy research "vanity publishing," while government officials in recipient countries have praised Bank research as helpful in their own conceptualization of relevant policy issues and priorities, even if it is unconnected to specific lending activities.[41]

Second, the most recent reorganization of the Bank hindered the ability of Bank staff to design stand-alone environmental projects. In the past, environmental staff in the Bank could visit countries and search for potential environmental projects. If they saw something that matched a country's environmental priorities, they could "raise interest" within the government.[42] Under the new system, country directors working with recipient governments on formulating their lending priorities are likely to find that environmental issues are relatively lower priorities for the latter than they are for the Bank's major shareholders. As a result, country directors find themselves calling less often on the environmental staff to design green projects, and conversely, task managers are more dependent on country directors for work. This results in a growing gap between those issues major Bank shareholders think are important for the Bank to pursue, and those preferred by recipient countries. Equally important, by treating the environment as its own sector, the latest reorganization has also reduced the incentives for staff to work across sectors, which means that the environment units have less ability to influence work in other sectors.[43]

Financial Tools

The financial tools available to an MDB are also an indicator of how banklike it is because they are designed for specific categories of recipients and activities. The fact that the IBRD is limited in its lending to government bodies has reinforced the Bank's interaction with policymakers and its emphasis on designing loans that have an impact on a country's broader policy priorities.[44] Sovereign-guaranteed loans are not only for individual investment projects, but for the more policy-oriented structural adjustment lending or sector lending as well. In the Bank's portfolio for individual countries, discrete investment projects, such as those that build infrastructure, will appear hand-in-hand with projects aimed at restructuring an entire sector, or developing new legal or regulatory frameworks.

While there is no set minimum loan amount at the World Bank, its projects tend to be larger than the EBRD's or the EIB's. In Central and Eastern Europe, for example, there are a large number of World Bank projects between $100 and $150 million, and the average project size is just under $100 million. This compares with an average size project of just under 20 million euro at the EBRD and about 50 million euro at the EIB. Given the Bank's top-down approach, and its ability to undertake broad sectoral restructuring, the Bank has more opportunities than do more bottom-up, demand-driven MDBs to shape policy dialogues with recipient countries, and to bring in environmental objectives if it believes they are important. Its tools are thus amenable to pursuing environmental activities, at least from the "supply side" of what the Bank can offer to a country.

In addition, World Bank staff have access to in-house sources of grant funds that can be used to finance background research, feasibility studies, technical assistance, training, and project preparation. This can give Bank staff the possibility of exploring more innovative types of projects or looking at particular issues in greater depth than they would be able to do otherwise.[45]

Yet even a top-down approach must contend with recipient demand in cases where the Bank competes with private sector banks and fellow MDBs for business in more economically advanced recipient states in the ECA region. The Bank's ability to sell environmental projects to countries where such activities are not a priority is greatly reduced when there are alternative sources of financing with fewer strings attached. In sum, the composition of the Bank's financial tools leave it well positioned to sell environmental projects and policy ideas to recipient governments, but it faces an increasingly competitive environment for doing so in more highly developed client states.

Procedures

Much of the literature on organizational theory begins with the assumption that organizational behavior is strongly shaped by the rules and standard operating procedures organizations follow. Yet organizations also struggle with the fact that these rules and procedures are not implemented as envisioned by the organization's designers. Rules may be revised and refined as the institution seeks to adapt to new challenges. Or, the gap between what the institution is supposed to do and what it is able to do may remain, without adjustment. The rules and procedures themselves say something about an MDB's efforts to address environmental issues, but even more important is how they work in practice; that is, how they shape or pose challenges to institutional incentives. In the case of the World Bank, the existence of numerous rules and procedures for addressing environmental issues provides further evidence of its development agency character vis-à-vis the other two MDBs. However, it is also clear that the World Bank faces a greater challenge in harmonizing these rules and procedures with its other development and banking goals.

The World Bank is well known for having an inordinate number of rules and procedures. The Bank itself recognizes that its "overly bureaucratic process" has reduced its ability to respond quickly in a changing external environment.[46] By the early 1990s there were over 200 different tasks that managers had to carry out in their work. These tasks, called "operational directives" (ODs), covered everything from environmental impact assessment procedures to the Bank's policy on resettlement.[47] Interviews with Bank staff revealed that they rarely paid attention to some of these ODs, because it was simply impossible to follow them all. Several ODs have been seen as weak or as failures and have been revised over time.[48] By 1993, the Bank began to reissue some of the major ODs as simpler, nonbinding "operational policies" (OPs), prompting debates on whether these would encourage greater laxness among project designers, or in fact would allow them to make projects less complicated and more easily implementable. Clearly the answer depends on which of the policies are being watered down and what impact a looser interpretation has in a specific project context. The following section turns to the major relevant rules and procedures as well as the additional factors that influence the Bank's porousness.

Country strategies   The project cycle is the overall framework that determines how an MDB is supposed to design projects and make loans, and each of its components contains its own rules and procedures. All MDBs have similar project cycle processes. These consist of: *project identification; project preparation,* where the bank and the borrower flesh out the details and specific conditions necessary to achieve the project's goals; *project appraisal,* where the bank undertakes a technical, economic, financial, and sometimes environmental review of the project to ensure it is soundly designed; *negotiations* between the bank and borrower agreeing to the specific loan conditions, including the economic, political, environmental, and other covenants built into the loan agreements; and *implementation and supervision,* where the borrower is responsible for implementing the project, with bank oversight.[49]

The World Bank's project cycle differs from those of the other two MDBs in the extent to which it situates individual projects within in-depth analyses it periodically undertakes of a country's economy. This provides the basis for developing the country assistance strategy (CAS), a key document between Bank and borrower that defines the Bank's framework of lending for the latter. The CAS is an important indicator of the Bank's nonbanking goals—that is, its desire to build a portfolio of projects out of an overall analysis of a country's economic and also political context. Bank staff discuss the CAS with recipient countries but are not obligated to change its conclusions in areas where there is dis-agreement.[50] The CAS document analyzes the country's economic and political position, lays out its development and economic objectives and prospects, and sets out the Bank's own strategy for the country along with the specific details of past and proposed assistance programs. For many years these documents were secret, not seen even by the Bank's board. Critics of the Bank argued that this allowed management to avoid scrutiny of its country strategies, while the Bank pointed out that many of its borrowers resisted more public knowledge of their economies and policies. Since the mid-1990s, however, as a result of the board pressure on the Bank and the Bank's movement toward greater openness, CAS preparation has become more transparent. The process now involves more active consultation with borrower countries, including members of civil society, and board consideration of CAS reports. In August 1998, the Bank's board amended Bank policy on CAS to allow the reports to be made publicly available, but only if the country agrees.

In theory, the CAS reports present an important opportunity for a country's environmental context to be discussed, because it can give Bank staff the ability to flag important environmental issues at the beginning of the policy negotiation process with recipient countries. Indeed, the national environmental action plans (NEAPs) that the Bank has required for IDA borrowers and suggested for IBRD borrowers are supposed to inform and be integrated into the CAS reports.[51] These NEAPs describe and analyze a country's major environmental problems and offer policy solutions and other actions for addressing them.

In practice, the CAS reports have done a poor job of integrating environmental issues into their conclusions. According to a 1996 internal World Bank review of twenty-five CAS reports, the links between a country's macro economic situation and its environmental issues were "not articulated."[52] The report criticized CAS reports for not addressing the impact of different development strategies on issues such as sound environmental and natural resource management. One reason for this is that the CAS reports must focus on a handful of development priorities, and environmental priorities are usually not competitive with respect to other priority issues in many countries, such as poverty reduction or macroeconomic stabilization. CAS reports also have a short time frame of one to five years, whereas the time horizon to address environmental problems is usually longer.[53] Coverage of environmental issues in CASs varies widely and often is accomplished either by addressing environment as a sector or through issues arising in other sectors, such as waste water treatment in the water sector.[54] The 1996 report recommended "some improvement in the Bank's organizational and incentive structure . . . to encourage wider and more effective consultation among experts, to reward team spirit, and to ensure collective accountability, so that a CAS can concisely state, for a country with multiple development goals, what are the key environmental challenges and how to design appropriate strategies, policies and action plans."[55] Significant changes have not been forthcoming, prompting an April 2000 Bank document to note that the environmental aspects of a CAS tend to remain isolated from the rest of the document, while attempts to link environmental issues to others discussed in the CAS remain weak.[56]

EA Procedures   "There is no question that the Bank's (environmental) procedures are exemplary. . . . What is controversial is how well it follows them," wrote the World Wildlife Fund's Katrina Brandon.[57] Indeed, the World Bank has a comprehensive set of rules for the ways in which individual projects must address environmental issues. These environmental assessment (EA) rules were the model on which the EBRD's own EA procedures were developed (in fact, by some of the same people), as well as those for some of the other regional development banks. They clearly distinguish the World Bank from private sector banks, and, in principle, can greatly shape project design. Compared with other MDBs, the World Bank's EA procedures have received the most scrutiny and have undergone the most revisions. They are also the most transparent and encourage the highest degree of public participation. Yet, integration of these rules into projects remains patchy, and the Bank continues to search for more ways to increase the effectiveness of EAs.[58]

EA procedures, brought into the Bank as a result of pressure from major shareholders as well as Conable, were codified in 1989 as an annex to Operational Directive 4.0 and subsequently revised and by 1999 converted to OP/BP 4.01.[59] The EA policy requires that all loans be screened for their potential environmental impacts. The purpose of the EA is to assess ways of preventing, mitigating, or minimizing possible adverse environmental impacts. EAs can be produced at the individual project level, as well as the regional and sectoral level, although they are not undertaken for structural adjustment loans, which have been as high as 63 percent of total lending.[60] EAs are formally the responsibility of the borrowing country, but Bank staff usually assist in preparing and monitoring them.

There are four main steps in the EA process. The first step is called *screening* and occurs in the project identification stage. At this stage, the task manager (with assistance from relevant environmental staff) identifies the magnitude of the project's environmental impact and assigns the project one of four categories. Those in Category A may have a significant impact on the environment and require a full EA. Examples of Category A projects include dams, power plants, large industrial plants, land clearance, and rural roads. Those in Category B have less significant impact and require a partial EA. Examples include small-scale irrigation and drainage, renewable energy, and tourism. Projects in Category C have no apparent environmental impact and do not require any EA; these include projects in education, nutrition, and telecommunications. A fourth category, "FI," was developed for projects that involve Bank loans to financial institutions, which in turn lend the money to subprojects that may have adverse environmental impacts.

In the *scoping* period, which occurs during the project preparation stage, the project's likely environmental impacts are further defined, and the task manager develops terms of reference for undertaking the EA. This is a point in the project cycle when local communities and NGOs are supposed to receive information about the project and become involved in consultations on it. In the third stage, an EA report is prepared and is reviewed during the Bank's appraisal. This report lays out the basic facts on the project, identifies its positive and negative environmental impacts, provides an analysis of alternative actions that can be taken to reduce the environmental impacts, and presents a plan that specifies measures to address the environmental issues. Finally, during the *project implementation* stage, the borrower is responsible for carrying out the recommendations of the EA report, and the Bank is responsible for supervising this implementation.

The Bank's EA procedures have evolved since they were first introduced, in part as a result of critical evaluations produced by the Bank's own internal reviews. For example, among the important changes distinguishing the 1991 version from the 1989 version are greater specification of requirements for public consultation, such as making draft EAs available to relevant groups, and a definition of environmental mitigation plans, which became part of the criteria for Category A and B EAs. The 1999 revision then brought sectoral adjustment loans under the umbrella of actions subject to EA requirements.

In theory, these procedures would ensure that the World Bank fills a basic condition of its environmental mandate—to try to avoid or mitigate environmental harm by controlling the environmentally destructive effects of its activities. However, EAs can have more far-reaching effects as well. They can have a fundamental impact on project design, for example, by forcing Bank staff to think more carefully about the environmental implications of the investments; by encouraging thinking and research on the risk assessment methodologies; by contributing to scientific knowledge on the flora, fauna, and general environmental conditions of a country; and by building environmental management capacity in recipient countries, as well as encouraging public participation in parts of the world unaccustomed to such trappings of civil society. EA procedures have the potential to be an important source of institutional change within the organization, while also having an impact on broader development issues.

Nonetheless, the Bank's EA procedures often fall short of their goals. An internal 1996 World Bank report assessing the Bank's EA procedures concluded that while staff generally comply with the procedures themselves, "very few EAs actually influence project design."[61] With regard to

Category A project in particular, the report was stinging: "The approach to environmental assessments used in the Category A projects reviewed by this study often generated massive documents that are of little use in project design and during implementation. Data collection efforts for EAs have not been adequately focused, which has added to the borrowers' costs without significantly improving the documents' usefulness. Bank terms of reference for EAs often required prediction and evaluation of too many impacts, obscuring the focus on relevant, key impacts."[62]

One of the main factors contributing to this problem is that the timing of EA preparation often does not fit well with the overall project cycle. An EA can often take more than a year to complete, which often means that by the time its conclusions are finalized, the project design itself is also finalized, "thus precluding meaningful consideration of alternatives," according to the report. This is even more likely to occur in cases where task managers see the EA as a "pro-forma ritual."[63] The report also criticized Category A EAs for only superficially considering alternative designs and technologies, as required. Even EAs that did consider alternatives "often explore weak, superficial or easily dismissed options." Another serious problem is the fact that often EAs are poorly implemented, in many cases because project implementation staff do not understand the EAs, or because the EA itself was not even available to the project office.[64] Instances where borrowing countries are not committed to accept or adopt the Bank's requirements also contribute to weak implementation.

Some of these problems are more easily correctable than others. Ensuring EA availability to project offices is relatively easy, for example, whereas harmonizing the EA process with the broader project cycle process might be more difficult. EAs that seriously consider alternative issues may be more thorough and expensive, but may recommend options that are not a priority for the recipient country and therefore are not implemented. As noted above, there are few incentives for Bank staff to promote time-consuming, expensive additions or changes to project design.

The ways in which the Bank's various environmental ODs and OPs are translated into projects, as well as the extent to which NEAPs have a real impact on domestic policy priorities, echo the Bank's experience with its EA procedures. For example, like EAs, NEAPs can strengthen the ways in which the Bank and individual recipients prioritize and address environmental issues, while also serving as an instrument to strengthen the capacity of domestic environmental ministries and as a forum for facilitating networking between domestic environmental actors and between these actors and the Bank. Yet evaluations of NEAPs by the Bank show that their quality and efficacy are mixed. Some NEAPs are more focused on specific priorities and projects than others, and even some of the more well-written NEAPs may be ignored by recipient country policymakers.[65] In some cases, recipients view the initiation of NEAPs simply as a way to comply with IDA requirements, rather than part of a broader process of policy reform.[66]

The Bank's experience with its environmental procedures and policies shows how innovative intentions often run into obstacles in practice. Conflicting or mismatched incentives are clearly a culprit, compounded by weak recipient government capacity or interest.

Porousness   The degree to which an MDB is open to the input and feedback of the public is also an indicator of how banklike and demand-driven the MDB is. Client confidentiality is sacred to private sector banks, which have no obligation to ask the public's advice on whether to make a loan and how to carry it out, and have a strict fiduciary responsibility to keep client-related data confidential. Public accountability and transparency, instead, are expected of public institutions. After decades of secrecy, the World Bank has recently transformed itself into the most transparent of the three MDBs. It was the first of the three to adopt a new information disclosure policy, in 1994, which made more Bank documents available to the public and created a public information center (PIC). The PIC makes available a variety of documents, many of which were previously confidential, such as staff appraisal reports (produced from 1994 onward), which describe projects and their implementation plans. Other available information includes economic and sector reports, environmental impact assessments, and project information documents.

Of the three MDBs, the World Bank has also put the greatest effort into building relations with NGOs. As mentioned in chapter 3, these changes were largely externally driven, and a result of United States pressure (backed by the threat of funding cuts) supported by NGOs. These efforts have included an NGO-Bank committee set up to expand relations and increase cooperation between NGOs and the Bank; a 1989 operational directive on "involving Nongovernmental Organizations in Bank Supported Activities"; and more recently, NGO sector studies in a variety of countries. By 1997, the Bank had placed NGO liaison staff in seventy-two of its resident missions and had launched a program to work with a global network of NGOs to review the impact of structural adjustment lending in a set of countries.[67] In Central and Eastern Europe, the Bank set up an ECA region NGO working group on the World Bank in 2000, to increase dialogue with civil society. NGOs are also becoming more involved in Bank project implementation and are occasionally involved in project design.

Another important indication of the Bank's "porousness" to outside stakeholders is its creation in 1993 of an independent Inspection Panel to investigate complaints from nonstate actors who feel the Bank has not conformed to its policies and procedures. The panel is the first mechanism that enables nonstate actors to hold international organizations directly accountable for their actions, and it has been a forum for nonstate actors who have challenged Bank actions in countries as diverse as Nepal, Brazil, Argentina, Bangladesh, India, Tanzania, and Ethiopia. The Bank has also become more openly self-critical in recent years, more commonly publishing reports that admit to past mistakes, an exercise not yet seen at the other two MDBs.[68]

In theory, the more open approach by the World Bank to interaction and dialogue with NGOs and its movement to make more information publicly available might be expected to lead to greater accountability and transparency, which in turn would impact the Bank's environmental work. In practice, however, the jury is still out, given the mixed response by NGOs and scholars to the Bank's activities in this area.

NGOs have conceded that the Bank's public information policies and relations with NGOs are far more developed than those found in other international institutions. Yet they also argue that many of the changes are not as far-reaching as they might appear to be. Although more documents are available to the public, many project documents developed before project approval continue to be confidential, which makes meaningful public participation in project planning difficult.[69] Areas of lending such as SAL remain relatively secretive. NGOs in developing countries have also complained that documents that are supposed to be publicly available continue to be difficult to procure.

The Bank's relations with NGOs, in turn, suffer from that fact that policies and intentions of engagement often clash with more pressing incentives in the Bank to move money while avoiding delays where possible. Bank staff have noted that public participation can be very expensive and time consuming, and they have admitted that many really have "no clue about how to do it."[70] Other staff, however, have gone out their way to involve the public in project design.

Paul Nelson, analyzing Bank-NGO relations, argued that the Bank's engagement with NGOs has "broken some new ground" among international institutions particularly in allowing "experiments and minor innovations" designed to encourage operational engagement.[71] However, the Bank's approach has been ad hoc, and ultimately it has not influenced its deeply rooted organizational characteristics such as the imperative to "move money." This means that often there is scant evidence that NGO participation in Bank-financed activities makes a significant difference.[72] Collaboration with NGOs, like other goals of the Bank, does not fit well with incentives and pressures for Bank staff to bring projects to the board in a timely fashion. Bank staff often see public participation as a necessary evil to placate the Bank's board, and therefore the quality of public participation in Bank projects can be low.[73] The picture that emerges from outside analysis of the Bank's policies on public information disclosure and NGO relations is that the policies are more far-reaching than those found at other IOs and MDBs, but implementation has been uneven owing to other pressures and incentives facing Bank staff. There is also criticism that the Bank's greater candor toward its own shortcomings has not translated into the Bank's being less tolerant of violations in its policies, given its perceived interest in not offending its major borrowers.[74]

The Inspection Panel has also had a mixed record in its activities and effectiveness. NGOs have credited it for being "one of the most important opportunities for improving the Bank's performance and accountability" and as a model for similar mechanisms created or proposed for other development banks.[75] The Panel's inspection of the alleged violations at the controversial Arun III project (mentioned in chapter 2) was cited by Wolfensohn as a reason for his having pulled the Bank out of that project. At the same time, NGOs argue that the Panel's efficacy is weakened by a lack of support for it from many members of the Bank's board of directors, who prefer to avoid embarrassing major client countries even when loan agreements are being violated. The Bank's process of streamlining many of its operational directives into less enforceable guidelines also may narrow the Panel's ability to rule on alleged violations. Finally, the Inspection Panel continues to lack autonomy and much power in practice, because it can investigate a complaint only if it receives approval from the Bank's board, and its recommendations are nonbinding.[76]

EBRD

The EBRD can be situated between the World Bank and the EIB in terms of the ways in which it has sought to institutionalize its environmental mandate. It is more banklike than the World Bank, with a narrower mission, fewer broad policy dictates, and less bureaucracy, and it prefers to pay more attention to funding specific projects than to addressing big policy reform issues. At the same time, it has gone further than the EIB in seeking to address environmental issues explicitly in its work by attention to both "add-on" and "build-in" techniques. In the first category, it has set up a specific unit to focus on energy efficiency projects, and its municipal department also emphasizes environmental issues and stand-alone projects in its largely public sector work. In the second category, the Bank's independent environmental appraisal unit has been able to exert influence on project design.

Mission

The EBRD's mission to promote economic transition in former communist countries, with an emphasis on private sector development, immediately made it somewhat difficult for Bank staff to pursue its early, ambitious goal of playing a "leadership role in the environmental recovery of the region."[77] The EBRD was never intended to be a "development" bank in the traditional sense, like other regional development banks that focus on Third World countries. Since it was not in the business of selling broad policy changes to recipient governments, it could be expected to have the most success in pursuing the environmental recovery of Central and Eastern Europe in areas that complemented its strengths as a private sector–oriented development bank, which largely meant infrastructure projects, including transport, energy, and municipal infrastructure. The EBRD's main activities are in areas such as strengthening the development of domestic financial institutions, supporting the privatization of state-owned industries, and mobilizing foreign investment in transport, telecommunications, and other categories of infrastructural restructuring. Many of these areas in fact entail lending to central governments, with sovereign-guarantees, but are seen as important in helping governments to develop the private sector.

The EBRD sees itself as driven by client demand—whether private or public sector—and is aware that its resources are limited compared with the scope of the economic problems facing the transition countries. As a result, stand-alone environmental projects are not a priority of the Bank except where such projects fit a country's own priorities, or overlap with other economic restructuring goals. The EBRD, then, cannot be expected to be a "darker green" MDB, toward the right end of the continuum presented in table 1.2. Yet, the Bank's mission still gives it the ability to undertake environmental due-diligence procedures, as a means of avoiding environmental harm in its work.

Structure, Location of Environmental Staff
With a staff of around 950, the EBRD is much smaller than the World Bank and has a much less complex organizational structure. The design of the Bank has evolved over its short life in ways that have resulted in dividing into discrete units staff focusing on environmental issues. The Bank underwent a significant reorganization when Jacques de Larosière became its second president in 1994. The reorganization merged separate divisions for merchant banking and development banking, and created one banking unit organized into northern and southern regions, made up of country teams and sector teams. These regions were in turn merged in early 1995. This evolved into a structure that consisted of three country groups (Central Europe, Russia and Central Asia, and Southern and Eastern Europe and the Caucasus) and three sector groups (financial institutions, infrastructure, and industry and commerce). The country team directors, in turn, would be located in the ten largest recipient country offices, to move more staff into the field. Projects are signed off by both country and sector staff. Operations support units include the Bank's appraisal unit, and the Bank's research is largely undertaken by the chief economist's office.

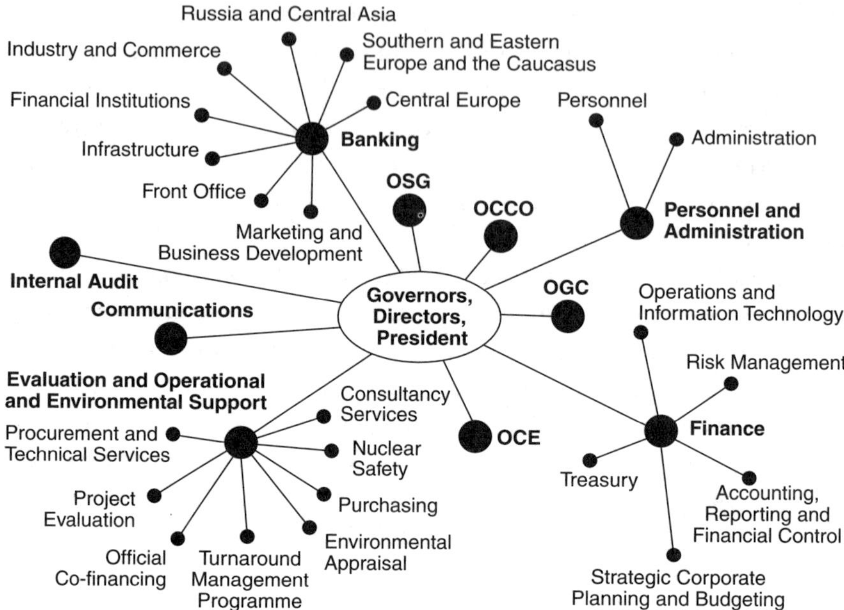

Figure 4.3
EBRD organizational chart. *Note:* OST—Office of the Secretary General;
OGC—Office of the General Counsel; OCE—Office of the Chief Economist;
OCCO—Office of the Chief Compliance Officer.

Bankers at the EBRD face the same basic incentives as those in the
World Bank and other MDBs to identify "bankable" projects and to get
them out as quickly as possible. An ideal bankable project is one that
generates a revenue stream (preferably in foreign exchange); and has a
significant local contribution; a sound institutional framework, and sufficient guarantee that the loan will be repaid. Many environmental projects are relatively less bankable because they do not generate short-term
financial returns and rarely earn foreign currency needed to repay MDB
loans denominated in hard currency. In addition, environmental projects
are often more time-consuming to develop than projects in other sectors
because of the need for intensive institution-building to make the project
viable; in the environmental sector, the banks are often dealing with weak
ministries, institutional structures, and problems not addressed in the past.

EBRD bankers also have incentives to finance bigger projects over
small projects, because loan volume is a factor in a banker's yearly perfor-

mance rating.[78] As a result, EBRD staff who do focus on environmental issues tend to be found in three particular areas of the Bank where they face extra incentives to do so: in its municipal and environmental infrastructure team, its energy efficiency unit, and its environmental appraisal unit. Under the broader context of the Bank's demand-driven mission, these particular units house the EBRD's "green bankers," whose job is to provide the substance behind the Bank's environmental mandate. Most of the Bank's environmental activities stem from these three groups. The first two units are operational units, and their staff seek to identify and design loans. The third has the opportunity to more deeply penetrate the Bank's operations, as its internal watchdog on environmental issues. In addition, the Bank has an environmental advisory council, an independent body of environmental specialists who meet twice a year and advise the Bank on environmental issues. It is also the secretariat for a nuclear safety account (NSA), a program initiated by the G-7 countries to offer Central and Eastern Europe and former Soviet countries financing for nuclear safety improvements.

The municipal and environmental infrastructure team finances projects in areas including water supply, sewerage and wastewater treatment, district heating, and municipal and industrial waste management. The team expanded significantly since it was set up—from around four bankers before 1994 (who were attached to the Bank's transport team), to a separate unit in 1995 that grew to around twenty bankers by the late 1990s. The team supports the decentralization of municipal services and infrastructure activity through a set of bankable projects that also have some positive environmental impact.

In some ways, the unit has behaved less like a merchant bank than other divisions in the EBRD, because it was funding mainly public sector projects that required sovereign guarantees—projects that would have been seen as too risky by private banks. As municipalities became more financially healthy, the unit has turned increasingly toward limited recourse or nonsovereign lending. The EBRD's municipal team tended to use more technical funds and bilateral grant money than other Bank units.[79] Before the 1994 Bank reorganization, the team was also more involved in policy-oriented work, such as research on environmental standards and public participation policies, but since then it has narrowed its focus to investments.

Indeed, some members of the Bank's senior management were averse to creating a separate municipal infrastructure team, on the grounds that "the merchant bankers didn't see environment and municipalities as important."[80] The board supported the idea of a separate municipal team, particularly the United States, the Nordic countries, and Austria.[81] The view was that the municipal sector was important in Central and Eastern Europe, since it had enormous needs and municipalities play an important role in people's everyday lives. Basic municipal services such as public transport, district heating, and water supply were seen as vitally important in the promotion of economic development. The needs of this sector were also tremendous; the EBRD estimated that more than 150 billion euro would be necessary in the municipal and environmental infrastructure sectors of its countries of operation to achieve basic service levels.[82] In many countries in the region, the system to finance municipalities collapsed with the end of the central government control, leaving crumbling infrastructures that were not able to supply basic human necessities like clean drinking water. It was also possible for the EBRD to find bankable projects at the municipal level, particularly since municipalities have some existing utilities (in water supply and wastewater treatment, for example) that can generate a cash flow from paying consumers. Although most municipalities were not strong enough to borrow money themselves (without a sovereign guarantee), the Bank financed projects that were expected to be able to pay for their own debt service. Municipal entities generally did not need to be built from scratch, but rather required restructuring and strategies and management that would allow them to perform efficiently while earning a profit. Said one EBRD banker, "There is enormous demand in the region. . . . Every town in Central and Eastern Europe needs investment in wastewater, solid waste and district heating."[83] Moreover, the team quickly developed a niche for its activities, working most actively of the three MDBs to develop innovative financing mechanisms and demonstration projects that included the use of municipal credit facilities, municipal guarantees, and an emphasis in helping municipal entities corporatize themselves in order to improve their performance and capacity to finance themselves.

The energy efficiency unit (EEU) at the EBRD was set up in 1995 to focus solely on financing small energy efficiency projects. The promotion of energy efficiency projects in CEE is important since former Communist countries have energy intensity levels between two and seven times average OECD levels. The EBRD's energy efficiency unit estimates that countries could save 30 to 40 percent of their energy consumption through demand-side and energy efficiency measures, which would equal almost the entire energy consumption of the United Kingdom and France together.[84]

The unit grew to around nine professionals by the late 1990s, and its work emphasized the development of energy service companies and credit facilities for projects that promote energy efficiency. It has focused on financing energy services companies (ESCOs). These tend to be small projects, where the Bank finances companies that in turn finance individual ESCO projects, mainly through multiproject facilities. In a typical project, the ESCO identifies energy savings opportunities in industrial, commercial, or municipal facilities. The ESCO makes changes in these facilities and is paid by its clients out of the energy savings achieved. The EEU also works on setting up energy-efficiency credit lines, whereby local financial intermediaries finance small- and medium-sized energy savings projects.[85]

The Bank's environmental appraisal unit (EAU) works as an internal watchdog to ensure that EBRD projects comply with the Bank's environmental policies and goals. This unit identifies all projects to assess potential environmental concerns and opportunities before projects are initially reviewed by the Bank's management committee, OpsCom (operations committee). If an EA or an environmental audit is required, the EAU helps the project leader draft terms of reference. It again reviews projects before they return to OpsCom for a final review ahead of board consideration. The EAU also undertakes a variety of technical assistance projects in areas that include legislation and standards development and has been involved in the development and revision of the Bank's environmental procedures.

Finally, the nuclear safety account (NSA) is a distinctly "un-banklike" activity, since it is financed by grant contributions from donor countries. The Bank administers the NSA, which was set up to undertake short-term improvements of Soviet RBMK and VVER 440/230 reactors as a prelude to their closure.[86] It also administers the Chernobyl Shelter Fund and specific funds supporting decommissioning in Bulgaria, Lithuania, and Slovakia.

Impact of Structure and Staffing on Green Activities

The structure of the EBRD has evolved to include two operational units that address environmental issues in their project work and an appraisal unit whose job is to ensure that the Bank's environmental policies and procedures are being followed. The staff in the first two units have special incentives to identify and finance projects with important environmental benefits, which means one would expect the EBRD's portfolio to include projects with significant environmental components or objectives. However, the "green bankers" in the municipal and energy efficiency units remain in discrete units, and some of the Bank's other units dwarf these "green" units in their output. By fiscal 2000, for example, the EBRD had financed a total of more than 1.5 billion euro worth of energy and power projects, compared with commitments of under 200 million euro by the energy efficiency unit.[87]

The EAU fills the gap to some extent, by addressing environmental issues for all projects, with the opportunity to insert environmental components into the design of projects and to build environmental components directly into loan covenants. The EAU, however, has no veto power when it comes to project approval. Like other support units, it can either sign off on a project or send a memo with reservations before the project is reviewed by OpsCom ahead of board consideration. The EAU is also part of the Bank's banking division, and to date it has not been known as a leading figure in pushing the Bank to find better ways of addressing environmental activities. The pressure for revised environmental policies and procedures, for example, was externally driven, mainly by the United States. In addition, NGOs and some ED members have long complained that the EAU has been weak in how it screens projects, placing projects that should have a full EA under the Bank's rules, into a category that does not require full EAs.[88] As a result, although the EAU is an important innovation in the Bank that pushes it to be more conscious of the environmental impact and design of its activities, there is some debate about how effectively its work is translated into action.

The existence of an environmental advisory council also appears to be an innovative part of the Bank's structure, but in practice the EBRD has not treated the council as an important resource. For example, the council's input was not sought when the Bank developed its first operational policies on energy and transport. Members have also noted that crowded agendas for brief meetings make it difficult for them to offer more than cursory contributions. Council members also have had little impact on the Bank's actual operations.[89]

Finally, the nuclear safety account is somewhat outside of the Bank's normal activities, since it is funded by grant contributions by donor countries, and projects prepared by Bank staff are submitted to the assembly of contributors. By the late 1990s, it had funded a handful of projects, which are discussed in greater depth in chapter 5.

Financial Tools

The EBRD's financial tools enhance its private sector orientation in ways that emphasize its character as a demand-driven financial institution versus a development agency and provide no special incentives for environmental lending that does not meet more important economic restructuring or financial goals. The Bank's emphasis on private sector lending gives it the flexibility of working without government guarantees in the majority of its lending and going directly to the private sector borrower. This means that government ministries and other actors are often not even aware of the details of specific EBRD loans, and these loans may or may not have an impact on broader government policy reform.[90]

The EBRD's tools themselves are more banklike than the IBRD's. In addition to making loans to clients, the EBRD can take minority equity stakes in projects, guarantee bond issues, develop credit lines, and provide venture capital through equity funds, among other activities. By the end of 1999, a total of just under 50 percent of EBRD financing was in the form of private sector loans, 29 percent in state loans, 22 percent in equity financing, and 1 percent in guarantees and other off-balance sheet items.[91]

The Bank also limits its financing for private sector projects to 35 percent of the total project costs and limits its minimum loan contribution to 5 million euro.[92] This provides great incentive for the Bank to cofinance big private sector projects with large Western companies, which then tend to reflect the demand for specific projects from foreign investors. Joint ventures in areas such as building Coca-Cola bottling facilities, improving ice cream production, and developing hotels are hardly the typical fare of development banks and are not areas of lending that emphasize environmental objectives.[93] At the same time, the Bank can argue that its activities are encouraging the economic development of Central and Eastern

Europe by energizing its private sector and attracting Western investors to enter markets they might otherwise feel are too risky. It can also argue that virtually any investment in new plants or equipment will be an environmental gain when compared with the old Soviet-style industrial capital stock. This way of promoting economic development, however, generally does not hold any particular incentives for pursuing projects with specific environmental goals other than where they coincide with other reasons for financing a project.

Procedures

The EBRD can be situated between the World Bank and the EIB in terms of the number of rules and procedures that must be followed in the project cycle. The overall project cycle at the EBRD is similar to those of the other two MDBs. While there are far fewer directives for EBRD bankers to follow than at the World Bank, the EBRD does have its own sectoral policies (in areas such as transport and energy, for example), which do not exist at the EIB. In the course of its existence, the EBRD has also strengthened its policies on the environment and developed new public information policies.

Country Strategies/Environmental Procedures

The EBRD does have country strategies, but these do not carry the same weight as the World Bank's country assistance strategies, because the EBRD is more demand-driven and less interested in policy-based lending. This means that the Bank does not engage in the type of policy dialogues with recipient countries that the World Bank undertakes. EBRD country strategy objectives tend to reflect the Bank's activities rather than lead it into new territory. They might include issues such as promoting private sector projects, strengthening a country's financial sector and infrastructure, and so on. Nonetheless, as a result of board pressure, the Bank's revised environmental policy also calls for country strategies to contain a section on the environmental implications of the Bank's work, which may draw on the World Bank's national environmental action plans for specific countries. This policy forces greater attention on environmental issues, but given that these country strategy documents do not drive lending, in itself it is not a strong incentive to shape how Bank staff identify and design individual projects.

The EBRD's environmental procedures are more similar to the World Bank's than the EIB's and were strengthened in a 1996 revision. Indeed, one of the key environmental experts who helped design the World Bank's EA procedures also helped to develop them for the EBRD.[94] The procedures are an important counterbalance to the Bank's demand-driven character, as a sort of "quality control" mechanism to incorporate environmental mitigation or enhancement into the Bank's operations.

Like the World Bank, the EBRD's appraisal process has a screening component whereby a project's potential environmental issues are identified, and a project may be placed in categories A, B, or C, which are similar to those of the World Bank. The EBRD may also undertake an environmental audit for manufacturing facilities or sites, to determine past or present concerns, risks, liabilities, and responsibilities for specific sites.

Under the Bank's first set of environmental policies, standards were chosen in a largely ad hoc manner, to fit particular projects. This does not mean that the chosen standards were low, but rather that there was no clear policy for determining which standards to use. Bankers preferred the flexibility to determine, on a case-by-case basis, which standards to follow, while environmentalists and some of the more powerful executive directors' offices (such as the United States and Germany, for example), felt a stronger policy was necessary to ensure that appropriate and high standards were being met.[95] "We thought they should have standards that you can quote," said an official in the U.S. executive director's office.[96] The environmental policy was tightened up in 1996, so that projects use EU standards for preaccession countries and meet national or World Bank standards where EU standards do not exist. For nonaccession countries, EU standards provide a common term of reference, and the Bank follows national or World Bank standards.

Porousness

The EBRD has not emphasized the development of relationships with NGOs and the public to degree that the World Bank has, but it has made significantly more efforts to improve its transparency than the EIB has. The Bank approved a new policy on information disclosure in 1996, in response to pressure by some powerful board members and despite opposition from some members of senior management.[97] The United States, in particular, made it clear that it would not participate in the Bank's first capital increase without this policy.[98] The underlying principle of the new policy is that information on the Bank's activities should be made available to the public "in the absence of a compelling reason for confidentiality."[99] This included the creation of new project summary documents, which provide information on both public and private sector projects before they have been approved by the board. The policy was revised again in 2000, adding issues such as the public availability of board-approved country strategies.

The description of the above policies shows some of the important ways in which the EBRD has sought to improve its environmental performance. Like the World Bank, the EBRD has a well-developed set of EA procedures and a relatively new public information policy. It does not have an independent inspection panel that seeks to ensure full compliance with its rules, and its country strategies do not carry the same weight as the World Bank's CAS reports.

As is the case with the World Bank, the EBRD's environmental procedures and public information procedures in theory provide incentives for the Bank to think more carefully about the environmental implications of its work and to avoid financing projects that might be environmentally destructive. Interviews with Bank staff and environmentalists provide a number of examples where project design has been enhanced by environmental consideration.[100] Although the EBRD's policies have not yet faced the type of scrutiny the World Bank has seen, they appear to face similar challenges and criticism; that is, the question of how well the Bank follows its own procedures. NGOs and major board members such as the United States believe the Bank's performance has been mixed.

Before the revision to environmental and public information policies, NGOs argued that the Bank often "cut corners" in implementing its procedures, or failed to implement them at all.[101] One case that stirred NGO criticism was a 1993 loan the EBRD made for investments in a new aluminum smelter in ziar nad Hronom, Slovakia. The EBRD's $110 million loan and $15 million equity contribution was designed to build a new smelter to replace the state-run ZSNP (Zavod Slovenskeho Narodneho Povstania) smelter that was built in the 1950s. The old smelter was one of the largest polluters in its region, with emissions of dust, SO_2, NOx, and carbon monoxide well above EU standards. It also contained what environmentalists called a "mountain" of bauxite waste, which was forty-five meters high and covered forty hectares. This waste leached arsenic, mercury, and cadmium into the soil and groundwater.[102] The project had already been rejected by the World Bank on economic grounds, given that Slovakia had to import all of the bauxite it made into aluminum and given the weak state of the world aluminum market.

Environmentalists were concerned about the plant and favored either its closure or its use for secondary aluminum production. What sparked their censure was their perception that the Bank did not follow its public participation procedures properly. As the Center for International Environmental Law (CIEL) noted, "There was no formal notification of the public; the public was not included in scoping the EA: meetings with a handful of environmentalists were conducted on extremely short notice and with no advance information provided."[103] The CIEL has also been critical of the Bank's EAs, arguing that they regularly lack some of the elements of a full environmental impact assessment, including an assessment of alternatives and indirect impacts.[104]

Even after the development of revised public information and environmental policies, NGOs continued to argue that implementation is uneven. They cited additional examples where the Bank did not follow its public participation procedures, including a Bulgarian Railway project and a proposed project to finance the completion of two partially built, Soviet designed nuclear reactors at Khmelnitsky 2 and Rovno 4 (K2/R4) in the Ukraine.[105] While NGOs criticize the Bank, they are also increasingly involved in the development of new EBRD policies, with opportunities to comment and offer feedback on policy drafts. NGOs are also in contact with EBRD resident offices in various recipient countries, although these offices do not have specific "NGO liaison" staff or environmental staff, as do some of the World Bank's offices in Central and Eastern Europe. By contrast, the EIB has no resident offices in Central and Eastern Europe. Some environmentalists have felt the meetings they have had with the EBRD Team Directors for their country as well as staff in the London headquarters have been positive, and they believe that access to Bank staff and documents has improved.[106] Generally, the EBRD appears to face the same struggles confronting the World Bank in this area. Although its procedures and rules call for more openness and public consultation, in practice it has been difficult to change a culture of bankers looking to avoid processes that slow down project development and further complicate the project cycle.

EIB

The case of the EIB is more clear-cut than the other two banks. The combination of less shareholder pressure to address environmental issues, along with the Bank's relatively more banklike mission, would lead one to expect to find less evidence of strong incentives for Bank staff to actively promote environmental issues. This expectation is borne out by the empirical evidence. Without the impetus provided by shareholder pressure seen in the other two MDBs, the Bank has attempted to meet its environmental goals primarily by emphasizing project adherence to European environmental standards inside the EU, and to relevant standards (EU, national, or international) outside of the EU. While the other two banks also incorporate national, EU, international, or World Bank standards into their projects, to a greater degree they also seek to fund projects with environmental objectives or major components. The World Bank in particular seeks to nest its activities in more specific policy analyses that may preclude some of the types of loans undertaken by the EIB.[107] In comparison with the other two banks, then, the EIB has not had to struggle as much to institutionalize its environmental mandate, because to date shareholders have not had high expectations for the Bank's activities in this area.

Mission

In terms of the Bank's mission, the EIB's dominant goal of promoting regional development inside its own shareholder EU countries explains the Bank's historical focus on traditional MDB infrastructure projects. Like the EBRD, the EIB can lend to both public and private sector projects. The emphasis on infrastructure lending was further encouraged in the 1990s when the Bank was asked by the European Council to focus more attention on the financing of trans-European networks (TENs), which consist of priority investments in transport, energy, and telecommunications projects that would encourage greater infrastructural links among member states (as well as preaccession countries). The Bank's other stated lending priorities include loans to small- and medium-sized enterprises, and loans to promote the international competitiveness of European industry.

EIB projects are not accompanied by the type of policy analysis and advice found at the World Bank, and to a lesser degree, the EBRD. The EIB has been called the EU's "house bank," or a "wholesale bank," despite its growing geographical scope, and it spends the vast majority of its resources and efforts in countries that are far more developed than the typical recipient of World Bank or EBRD loans. As a result, from a policy perspective, the EIB's primary mission has not evolved in line with the geographical scope of its lending.

In Central and Eastern Europe, it is somewhat ironic that the EIB has been the least policy-oriented of the three MDBs, because it is poised to become the most important MDB in CEE, particularly for countries joining the EU. As CEE countries build more healthy, developed market economies, they gain increasing access to private sector sources of finance, including international capital markets, so their appetite will decrease for relatively more expensive MDB loans that come with conditionality. By the mid-1990s, the World Bank and EBRD began to move away from the more developed countries in the region and started to focus greater attention on the less developed countries. Both banks also have graduation policies, whereby they use certain criteria to determine when a recipient country is no longer eligible for their loans. By contrast, EIB lending to CEE has increased, and will continue to do so as the EIB takes on a more prominent role within accession countries.[108] For the period 2000–2006, as an example, the Bank is authorized to lend 8.7 billion euro to the ten applicant countries, as well as Albania, Bosnia-Herzegovina, and FYR Macedonia, plus provide an additional 8.5 billion euro in a new preaccession facility for candidate countries for the period 2000–2003.

Structure, Location of Environmental Staff

The EIB's organizational structure offers limited incentives for Bank staff to pursue environmentally oriented projects beyond the more typical infrastructure projects that have positive environmental affects. One important reason is that the Bank is a lean operation, with its staff of around 1000 now lending more each year than the World Bank does with a staff over nine times that size. By the late 1990s EIB staff were organized into

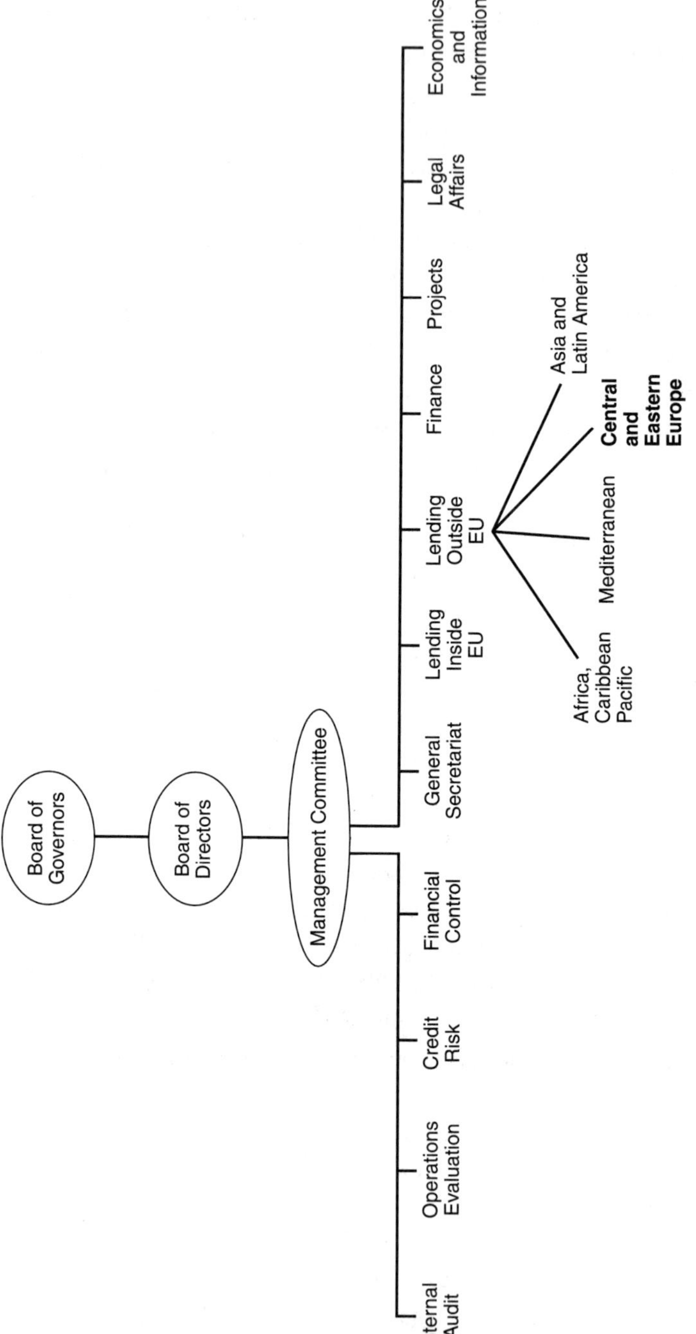

Figure 4.4
EIB organizational chart, 1999.

seven directorates, out of which one handled lending within the EU, and one covered all handles lending outside the EU. Within the latter directorate, there is a department that works on Central and Eastern Europe. This department contains approximately twenty bankers who cover thirteen countries. The fact that there are only around two bankers working on each country in the region has encouraged these bankers to rely on potential loan recipients coming to them with ideas, as well as encouraged the EIB to piggy-back onto projects developed by one of the other two MDBs as a cofinancier.

Given that the Bank is largely demand-driven, potential loan recipients (called "project promoters") tend to bring their ideas to the Bank. If the Bank chooses to pursue a project idea, appraisal is undertaken by a team consisting of a loan officer from one of the lending directorates and an economist and engineer from the Bank's project directorate. Appraisal teams can also include a lawyer from the Bank's legal department. The project directorate is a product of the 1995 merger of two separate directorates (one containing economists, the other containing engineers), which is divided into departments by sector (e.g., industry, energy and mining, infrastructure) rather than by country. The project directorate produces the appraisal report, which is used by the Bank's management committee to decide whether to recommend loan approval by the Bank's board.

The EIB has no separate environmental appraisal unit and no specialized bankers or division whose work emphasizes finding projects with primary environmental goals or significant environmental objectives. There is also no real research arm in the Bank, although a small chief economist's office was created in the mid-1990s to think specifically about broader strategic issues affecting the Bank. There is an "environmental policy coordinator," a position in the projects directorate created by the Bank in 1996 to help develop and ensure implementation of the Bank's broad environmental policy and to represent the Bank in various environmental fora.

As a result of this organizational structure, there is no particular community of environmentally oriented staff at the EIB, in the sense that there is at the other two banks. There are individuals who address environmental issues as part of their work, but these issues are not driving their work or encouraging specific staff to identify particular types of projects or policy issues. The engineers in the Bank's project directorate end up being the closest thing the EIB has to "green bankers," since they address environmental standards in their technical appraisal of a project, which is discussed in greater depth below. A few of the Bank's economists are environmental economists and do get involved in the Bank's work on environmental issues. Staff in the Bank's legal affairs directorate also work on ensuring the Bank is complying with Community law on environmental and other issues. However, the EIB does not have the types of in-house environmental networks that one finds at the World Bank and to a lesser extent at the EBRD. The EIB's lean staff ultimately precludes much involvement on a wide array of nonbanking issues. As a March 2000 report by Britain's Department for International Development concluded, "The Bank is not adequately staffed to assess all aspects of development soundness."[109]

Financial Tools
The EIB's financial tools enhance staff incentives to rely on projects being brought to the Bank by potential borrowers, as well as opportunities to cofinance projects with other MDBs. The Bank does not generally fund more than 50 percent of a loan project's total cost, so it relies on projects where there are partners. Given its small staff size, the EIB has been attracted to cofinancing projects in Central and Eastern Europe with the other two MDBs, and in such cases, the EIB does not take the lead in project development. Instead, it funds a discrete part of a project, and tends to rely on the other two banks for the bulk of the necessary work in project development and missions. The EIB also does not have the in-house sources of grant funding available at the other two banks to finance feasibility studies and background research for project development.

Although it is more banklike than the other two MDBs, the EIB ironically is also the most risk-averse and seeks double guarantees (both sovereign and EU guarantees) for most of its loans outside of the EU.[110] This also creates incentives for staff to look for more traditional infrastructure projects, which the EIB has historically funded, and provides a disincentive for innovation in the scope of work the Bank has traditionally undertaken, where that work may entail new or different risks. It would be harder, for example, for the EIB to find novel ways of undertaking municipal financing in CEE where sovereign guarantees are no longer forthcoming.

In Central and Eastern Europe, the EIB offers the cheapest loans of the three MDBs, with the least attached conditionality. Because it does not undertake the type of policy work and feasibility studies undertaken by the other two banks, it tends to have lower administrative costs. EIB loan rates are generally closer to LIBOR rates than those charged by the other two banks, and the EIB also does not charge a commitment fee on its loans, which the other two banks charge. This fee, which can be around 1–1.5 percent of a loan, is the MDB's commitment to reserve capital once a project is signed. In addition to its direct project loans, the EIB has been active in the use of "global loans," which are similar to credit lines that are set up for financial intermediaries to on-lend for smaller projects. This is typically a good way for the MDBs to be able to reach small- and medium-sized enterprises. The use of global loans also enhances the Bank's demand-driven character, since it does not know in advance how the resulting loans will be spent. Financial intermediaries have no particular incentive to seek out environmental loans or projects.

General Procedures, Environmental Procedures

Of the three MDBs, the EIB has the smallest set of bureaucratic rules and procedures governing project development. It does not have country strategies, and given its demand-driven nature, it often does not even design or develop the projects it lends money to. Project promoters present projects ideas to the Bank, and the Bank will evaluate projects of interest through an appraisal procedure. Because the Bank often enters into projects that are already developed, it can be more difficult for it to insist on changes to project design.

Environmental issues are addressed as a part of the broader routine of investment appraisals, undertaken by the project's small appraisal team. The purpose of the appraisal process is to evaluate whether a proposed project complies with relevant European and/or national laws and other rules, and to evaluate the project's economic, technical, and financial characteristics. Generally, it is the engineers who address environmental issues in the appraisal report. Their emphasis is on ensuring that projects within the EU comply with various Community legislation or national legislation—whichever is higher—and that non-EU projects comply with relevant environmental legislation. Outside the EU, the Bank has more flexibility on which standards to follow. Often it is up to the EIB's engineers to decide which standards to follow, and there are no specific guidelines at the Bank on appraisal. "Relevant" legislation and standards can be defined as EU (in the case of accession countries), national, or in terms of norms of "best existing practice." This often means making sure that the technology used in projects is likely to fulfill certain standards.

Within the EU, whether or not a project is subjected to a formal environmental impact assessment (EIA) is determined by the Commission's directive on EIA (85/337), which in 1997 was amended by directive (91/11). This directive serves as a reference for the project directorate's appraisal. Increasingly, this directive has become the norm for guidance in pre-accession CEE countries, although whether inside or outside the EU, it is the project promoter—the client—who is responsible for carrying out the EIA. The EIA directive identifies types of public and private sector investments that require an EIA. EIAs are undertaken by developers and must be provided to public authorities, who take the information into account in deciding whether to authorize the investments. The directive is mandatory for large projects such as building highways, railway lines, oil refineries, and large power plants, which are called Annex 1 projects and require public consultation.[111] And an additional set of projects (Annex 2) requires impact studies if the member state believes it is necessary.[112] The information required under the EIA directive includes basic information on the project and the ways in which it might affect the environment, alternative solutions to reducing this impact and particular measures to avoid or reduce potential negative impacts.

The ways in which environmental procedures have been institutionalized at the EIB create incentives for staff to rely heavily on pursuing relevant environmental standards in project design versus other ways of evaluating environmental options, including broader macroeconomic and sectoral analyses that could impact the strategic directions that lending might take.[113] Such a reliance on standards can produce the curious outcome where the EIB is using acceptable technology or standards in projects that might not make environmental or economic sense in a broader analysis.

The Bank's emphasis on standards works best when the appraisal team is working on EU-based projects in areas where there are strong, clear environmental standards, whether EU or national, or when engineers emphasize the use of technology that encourages energy or natural resource conservation (at a reasonable price) that might not have been addressed otherwise. Yet it can be problematic to apply EU standards outside the EU, where they may not always be appropriate. Reliance on national environmental standards in non-EU countries where compliance and enforcement may be weak can also water down the EIB's standard-based approach.

EU environmental policy itself is weak in several respects. Many areas covered under national environmental policies are not yet addressed at the European level, particular in the area of industry-specific standards. In addition, EU environmental policy uses directives as the primary legislative instrument, but these are not directly translated into domestic law. Member states determine the forms and methods of application. Since member states' record on implementing many environmental objectives remains poor, these environmental directives are often criticized as weak.[114] The EIA directive itself has been criticized by a number of sources for its mixed record in terms of implementation. The European Commission itself, for example, has argued that many member states abuse the discretionary power given to them under the Annex 2 types of projects.[115]

In addition, and more important, the EIA directive is a *procedural* directive. This can have a positive impact on areas where developers and planning authorities are compelled to address environmental issues that might not have been considered otherwise, but it may be much weaker in countries where there might not be the administrative capacity necessary for this procedure to be properly undertaken. While a number of CEE countries have adopted new EIA policies, in many of these countries these policies are unevenly implemented or ignored.[116] Norms for EIA and public participation continue to evolve in a region where governments are still learning how to address openness and transparency in decision-making processes. Since the EIA process is the responsibility of the government, there is little the EIB can do if a CEE government does not have the ability (or interest) to pursue EIA work.[117]

The way the EIA process has been institutionalized at the EIB has been criticized as too relaxed by some Commission officials, even for EU projects. Under Article 21 of the Bank's statute, for proposed loans in EU member states the Commission and relevant member state provide their opinions on the project before it is submitted to the Bank's board. The

Commission's vote on the EIB's board has particular strength; if the Commission votes against a project, the rest of the board must be unanimously in favor of the project for it to be approved.

DGXI, the Commission's environmental directorate, has felt that information circulated by the EIB related to EIAs has often been insufficient. For Annex 1 projects, for example, the Commission requires the EIB to "satisfy itself that the procedures . . . are made."[118] The Commission may ask for a nontechnical summary of the EIA before the EIB board approves the loan. For Annex 2 projects, and projects not falling under the EIA directive, the Commission still requires the Bank to assure it that the project is "acceptable" in environmental terms, in order for the Commission to formulate a positive opinion. DGXI officials have said that these procedures still allow for the EIB to provide the Commission with very little information, which sometimes makes it difficult for the DGXI to approve a project.[119] The EIB, in turn, believes DGXI should have more faith in the EIB's technical expertise.[120]

Porousness

For much of its existence, the EIB has not faced strong pressure to be more transparent. Only in the late 1990s were NGOs in Europe and CEE first beginning to organize campaigns to push more strongly for greater transparency. Of the three MDBs, the EIB is most resistant to opening up channels of communication with NGOs and other outside actors. The Bank has organized several meetings with NGOs, but relations have remained prickly. Friends of the Earth noted in 1997 that these exchanges "have yet to yield any substantial results in terms of changed policies and procedures."[121] Other EU institutions were also beginning to grumble about the Bank's lack of transparency. A March 2000 report to the European Parliament's Committee on Budgetary Control for Financial Activities assessing working methods and control mechanisms in the European Central Bank, EBRD, and EIB said the committee received full cooperation from the first two, and "stonewalling" from the EIB.[122] The Bank has also been accused of refusing to accept more scrutiny by the EU's fraud unit.[123]

The Bank first adopted rules on information disclosure in 1997, a year after the EBRD adopted its new policy and three years after the World Bank did the same. The EIB's rules are by far the least open. They do not specify any particular project documents automatically be available to the public (such as, for example, the project summary documents and EIAs available to the public from the EBRD, and a multitude of project documents and information available from the World Bank's public information center). Whereas the information policies of the other two banks stress circumstances under which information *will* be made available, the EIB's policy stresses all the reasons why it will *not* provide information. These include whenever, "in the judgment of the Bank . . . (a situation) is of such a nature that its disclosure could harm the Bank's legitimate interests." Examples of when this might occur are not defined. According to the EIB's rules on information disclosure, the Bank "has to respect, in its relationships with other institutions, bodies and persons both within and outside the Community framework, the confidentiality of communications and agreements between itself and the latter."[124]

The ways in which the EIB has institutionalized its environmental goals reflect a lack of real pressure from its board on environmental policy issues. Given its emphasis on following environmental standards versus using a broader array of analytical tools, its small staff size, and the absence of a stand-alone environmental appraisal unit or any other environmentally oriented home for "green" staff, one would predict that any environmentally oriented work in Central and Eastern Europe would stem either from the Bank's traditional activities in infrastructure lending or from cofinancing opportunities with one of the other two MDBs. Chapter 5 shows this to be the case.

Conclusion

The chapter shows how diffuse governance structures of the three banks can make it difficult for donor policy goals to be translated into staff incentives. Even in cases where donors have given an MDB a strong environmental mandate, its ability to institutionalize environmental policy goals is clearly shaped by the degree to which it is designed to emphasize its character as a financial institution versus a development agency. In particular, the incentives facing staff to design environmentally oriented projects or to incorporate environmental components into projects often do not complement other incentives to get bankable projects approved by the board of directors without what some bankers may see as undue delay. Stated differently, the incentives for staff to address issues of sustainable development often do not match other incentives to move money.

Where an MDB begins with limited environmental policy goals, we would not expect institutional design to provide many incentives for Bank staff to pursue environmentally oriented work. This chapter has shown this to be the case with the EIB. However, in cases where the MDB has strong environmental policy goals, such as the World Bank and EBRD, the incentives facing staff to translate these goals into projects and programs depend on how banklike the MDB is. The chapter has also shown areas where MDBs struggle to harmonize conflicting incentives. The role that the banks' organizational environment play in shaping institutional behavior highlights some of the limits of neorealist and neoliberal institutionalist theories.

Given the way the banks' environmental policies are institutionalized, what types of outcomes might we expect to see in their activities in Central and Eastern Europe? First, this chapter shows that "green bankers"—MDB staff whose job specifically requires them to focus on environmental issues—are located in specific departments in the World Bank and EBRD and consist of one official at the EIB. One might predict that World Bank and EBRD projects with environmental goals or significant environmental components could be traced to these people. A parallel prediction would be that such projects are unlikely to be found in the EIB's portfolio, unless such environmental goals coincided with other economic or financial goals and/or were demanded by project promoters.

Second, in general, it is likely that the relatively more demand-driven MDBs will emphasize as "environmental" those projects that fit their financial and economic criteria that are demanded by governments or by private investors and that have some environmental benefit. These might include energy rehabilitation projects, or water and wastewater treatment projects, as well as waste management, among others. The "green bankers" in the MDBs would likely be the source for projects that go beyond these categories, to include projects that are less demand-driven, and the incorporation of environmental components into projects that do not emphasize environmental issues.

Third, and following from the previous two points, we would expect to see a greater emphasis on projects with major environmental goals or significant environmental components in the World Bank's activities in Central and Eastern Europe, followed by the EBRD. The World Bank is designed to have the "darkest green" portfolio, whereas the EBRD has a "medium green" approach of an MDB seeking to address environmental issues without moving too far beyond its emphasis on private sector development, despite the ambitious environmental mandate it received from its shareholders. We would expect the EIB's portfolio to have the fewest of these types of projects, reflecting the absence of a strong mandate from its shareholders and reinforced by its institutional design. The following chapter examines these hypotheses in light of evidence of the three banks' activities in Central and Eastern Europe.

5

MDB Environmental Policies and Practice in Central and Eastern Europe

In postcommunist Central and Eastern Europe, the three multilateral development banks faced a dual challenge of how to help promote fundamental economic restructuring in a region that also contained some of the most polluted areas in Europe. This chapter analyzes how the three banks have addressed environmental issues in their lending and nonlending activities in CEE. The evidence supports the hypotheses suggested at the end of chapter 4. The banks' overall portfolios do reflect how bank-like they are, with the World Bank undertaking the widest scope of activities and projects with significant environmental goals and components, followed by the EBRD, and with the EIB trailing in a distant third place.[1] Further, the environmental character of the banks' portfolios reflects the activities and output of "green bankers." These institutional factors, and the incentives they influence, also highlight some of the gaps between mandate and actions, particularly for MDBs like the World Bank and EBRD that have strong environmental policy goals. The chapter also demonstrates more explicitly the ways in which recipient demand and other domestic-level factors influence the scope and performance of MDB activities.

The chapter is divided into two sections. It first situates the banks' activities in CEE, by describing the environmental issues facing the region since the fall of the Iron Curtain, and the ways in which major donor agencies have responded through environmental assistance strategies.[2] The three banks are the major international financial institutions operating in CEE, together committing close to $40 billion in 1990–1999.[3] The chapter then turns to the individual banks, examining their general strategies and activities in the region before describing and assessing specific environmental activities and loans. Environmental loans include those with primary environmental goals or significant environmental objectives. For a window on implementation issues, I focus on the performance of "green" projects in the water and energy sectors. These are common and important sectors of lending for all three banks, sectors where project decisions have obvious impacts on the environment—both positive and negative—and sectors the banks all stress as illuminating their environmental lending. In fact, energy-related projects comprise the largest single category of the banks' "environmental" loans, in terms of numbers of projects and lending volume. Water-related projects are the second largest category in terms of numbers of projects; they are also second in terms of loan volume for the EIB and EBRD (third for the World Bank, behind forestry projects). Project performance in these sectors also highlights many of the challenges facing the banks, showing more clearly the gaps between intention and practice, and the difficulty of measuring the "environmental impact" of MDB projects.

The Environmental Context in Central and Eastern Europe

The collapse of communism in Central and Eastern Europe revealed enormous environmental degradation. At the time, reports from the region drew public attention to dramatic issues such as the catastrophic destruction of forests by acid precipitation, high rates of pollution-induced sickness, and "crisis" proportions of water pollution, as seen in areas where water was undrinkable and entire rivers were dead.[4]

The compilation of data soon showed that much of the environmental damage in the region was located in specific areas. In Poland, for example, while some parts of the country were relatively unspoiled, around one-third of the population lived in 11 percent of the country's territory considered to be an "ecological hazard" due to a concentration of heavy industry and mining.[5] One of the cities in this area was Katowice, where according to a 1991 World Bank report, 24-hour ambient concentrations of black smoke were more than six times higher than European standards.[6] Katowice was part of a broader, regional "hot spot" called the "Black Triangle," encompassing many highly polluted areas in southwestern Poland, northern Bohemia in the Czech Republic, and southeastern Germany. In northern Bohemia, for example, large coal-burning power plants contributed to SO_2 emissions that measured 20 times the national average in the early 1990s.[7]

Elsewhere in the region, there were a number of small, old industrial areas that exposed nearby populations to harmful emissions. These included places like Copsa Mica, Romania, where two lead smelters and other badly maintained industrial facilities were thought to be the source of respiratory problems and high lead exposure;[8] and Dimitrovgrad, Bulgaria, where high hydrogen fluoride and hydrogen sulfide emissions from one poorly located fertilizer plant were thought to be a major source of below-normal height, weight, and lung function in half of Dimitrovgrad's children.[9]

Massive air pollution, which produces particulates and SO_2, is widely seen as the worst pollution problem in many parts of CEE in terms of its impact on human health. It is produced largely by the burning of domestic hard and soft brown coal as a major fuel. Soft brown coal, or lignite, is more prevalent than hard coal in the region and is characterized by its relatively high sulfur and ash content, as well as its low heating efficiency.[10] Major sources of air pollution in the region include power plants, small industry, households, and district heating plants that rely on coal. When CEE countries embarked on the journey of political and economic transition to market economies, they had the distinction of producing over 66 percent of Europe's sulfur dioxide emissions, but only 33 percent of its GDP.[11] The health effects resulting from coal burning in the region include high rates of chronic bronchitis and asthma, as well as other acute and chronic respiratory diseases.[12]

Water pollution has been another serious problem in the region. Data have shown that an unusually high proportion of surface water is not fit for human consumption. In Poland, this figure was put as high as 96 percent in the early 1990s.[13] Czechoslovak government statistics for 1992 showed that around 50 percent of the country's drinking water was below government purity standards, while 30 percent of the country's river water was not even capable of sustaining fish.[14] More recent data from the Dobříš Assessment showed one-third of the river water in 10 countries in the region had quality levels that were "bad" or "poor."[15] In addition, in many countries in the region, only a small percentage of the population was served by wastewater treatment facilities that treat wastewater to secondary levels.[16]

Water quality can threaten human health through the presence of nitrates (particularly in rural areas, owing to agricultural run-off) and the discharge of heavy metals and toxic chemicals. The main causes of water pollution in the region have included: inadequate drinking water processing systems and sewage treatment; unmonitored municipal and industrial discharges; contamination from agricultural runoff; contamination from industrial air emissions; and in coal producing regions, discharges from coal mining operations. Finally, for both water and air pollution in the region, at the time of the regime changes monitoring was mostly inadequate and technology was out of date. Additional environmental problems that have threatened health in the region stem from lead contamination in soil and air due to smelting and transport, and from inadequate disposal of toxic waste.[17]

Views on the Causes

Strategies of economic development adopted by the former, centrally planned regimes are widely blamed for the degree of Central and Eastern Europe's environmental degradation. The Stalinist path of economic development focused on state ownership of the means of production and favored rapid industrialization with an emphasis on heavy, resource-intensive industry such as chemicals and steel. Growth was the governing priority, with little awareness of the significance of environmental protection. Prices were determined by the state, not the market, and the prices of energy and other natural resources were often set at extremely low levels, which encouraged industry's excessive and wasteful use of energy. Central plans in the region did not include any mention of environmental protection until the 1970s, and even then there were few efforts to address growing pollution problems.[18]

Another factor contributing to the level of environmental degradation in the region was the absence of open public debate or opposition to government policy before the 1970s. In most countries in the region, individuals had no means to oppose state action and were not allowed access to environmental information, the main weapon of NGOs in any political system.[19] Often, environmental data were considered state secrets and were suppressed.[20] Curiously, many countries in the region did have strict environmental standards, but the environmental ministries had little power to enforce them.

Environmental Reform Process

Environmental issues began to be discussed in depth throughout the region during the 1970s, as environmental degradation became more visible at the same time that governments sought legitimacy by attempting to improve living standards and allowing some areas of public opposition to government policies. By the late 1980s, the environment had become a rallying point for people who were seeking broad political and economic change. In Hungary, for example, the proposed Gabcikovo-Nagymaros hydroelectric dam on the Danube triggered the creation of the country's first environmental movement, one that became a powerful voice calling for the end of the communist regime.[21] Public criticism of the dam project was synonymous with public criticism of the government and calls for greater political participation.

In the wake of the demise of the communist regimes beginning in 1989, environmental clean-up and protection emerged as one of the top policy priorities throughout the region. There were great hopes both inside and outside government that the region's environmental legacy would now be addressed openly, and that this was a historic opportunity to incorporate environmental protection into the new economic order. The environmental lobby organized around the Gabcikovo-Nagymaros dam dispute, for example, immediately achieved its goals when the newly elected Hungarian government suspended construction on the Hungarian side.

New environmental ministries were set up, or old ones restructured, and they were given the mandate of establishing basic environmental priorities. Most countries in the region have now enacted framework legislation on the environment, which contain principles such as polluter pays, prevention and precautionary action, and public participation. Countries are also in the process of establishing or revising environmental standards, increasingly with the goal of bringing them in line with European standards. A number of CEE countries have established environmental funds, which collect environmental fees and fines that are used to finance grants or low-interest loans for environmental purposes. Yet, as economic restructuring goals assumed prominence, the initial enthusiasm toward environmental restructuring and reform faded.

The trend changed again in the late 1990s among countries preparing to join the EU, since meeting EU environmental requirements (the *acquis* or the body of directives and regulations that must be met) requires these countries to undertake an enormous level of infrastructural investment and legislative reform. Negotiations formally began with the Czech Republic, Estonia, Hungary, Poland, and Slovenia in December 1999 on the environmental chapter of the EU Treaties, with the other accession countries hoping to follow. The European Commission estimated a cost of around $120 billion for the ten CEE accession countries to bring their environmental standards up to European norms.[22]

Progress on environmental policy reform and clean-up remains mixed. While governments have initiated numerous new policies and set up stronger legal frameworks, environmental management institutions in many countries in the region remain weak. This reflects the lack of sufficient financial resources available to most of these ministries, relatively weak administrative capacity, and the fact that environmental ministries tend to be among the least politically powerful ministries. New environmental problems have also cropped up, reflecting some of the changes brought about by the transition process—a vast increase in the number of automobiles in the region, and new pollutants (such as detergents) entering the water supply without treatment.[23] By the late 1990s, because none of the accession countries had made much progress in applying EU environmental laws, EU Commission President Romani Prodi called for less pressure on the CEE countries to meet EU environmental levels before joining the Union.[24]

At the same time, in many parts of the region some important pollution problems have declined. For example, air quality has improved substantially in many areas, reflecting a downturn in industrial production, higher energy prices, and the restructuring, modernization, or closure of some old industrial polluters. Many power plants throughout the region have now made the switch to clean natural gas, while the use of scrubbers at coal-burning plants has increased. Some countries, such as Slovakia, the Czech Republic, Hungary, and Poland, have made significant progress in reducing the use of leaded gasoline.[25] In Poland and Hungary there are even signs that a return to economic growth has not resulted in increased sulphur dioxide and carbon dioxide emissions.[26] The quality of the region's rivers and lakes has also slowly improved in some areas where industrial discharges have fallen, although residue from previous emissions can remain a source of pollution.[27]

Role of Donor Institutions: Regional Overview

The activities of MDBs are an important component of the broader landscape of donor environmental assistance to the region. As environmental concerns became a relatively less important policy priority in CEE, donor institutions acted as a powerful countervailing force to catalyze environmental reform and clean up in the region. Bilateral and multilateral donors have influenced policy development by altering the costs of options available to domestic policymakers through the provision of policy and project ideas backed by technical expertise and financial resources in domestic contexts characterized by weak institutional capacity and insufficient financial resources.[28] Donors influenced the policy development process directly, by providing resources and ideas for specific environmental projects that governments would be unlikely to fund on their own. Indirectly, these actors helped to shape the parameters of policy debates, assisted with legislation development, and supported environmental interest groups.

Donors and recipients knew fairly quickly that the costs of cleaning up various environmental problems in the region would be enormous, and it was evident that donors would contribute only a small portion of the total cost. For example, the World Bank calculated that for Poland alone the cost of meeting EU environmental rules would be at least $6–13 billion a year for the next fifteen years. This could be doubled, if the EU were to tighten and strictly enforce its standards.[29]

Totals for donor aid are dramatically smaller than cost estimates. There is no source of precise data that provides aggregate amounts of environmental aid. The only current source is compiled by the OECD, but it does not show its methodology for deriving its figures, which seem to be on the low end.[30] It calculates that between 1991 and 1997, major bilateral donors and the three MDBs committed around 2.5 billion ECU (or approximately $2.7 billion) for investment in projects with primary or significant environmental components, as well as policy development and technical assistance.[31] My own data, analyzed later in this chapter, shows that the three MDBs alone committed around $2.9 billion in financing for projects with significant or primary environmental goals through 1999, and this figure does not include money spent on technical assistance.[32] The point is that there is an enormous gap between financing needs and assistance levels. Given this gap, donors have realized that assistance must be leveraged to be effective, and since 1991 they have made conscious attempts to work more closely together and with recipient country actors to address the region's environmental problems. Donors have also sought to use aid to improve environmental management in the region and to encourage a long-term process of environmental reform.

In 1991, what is now referred to as the "Environment for Europe" process was launched in response to early failures in aid coordination and prioritization.[33] These early failures were characterized by numerous feasibility studies that did not lead to investments, low absorption rates for assistance, or duplication and overlap in donor activities. Donor environmental assistance has been driven by factors that include concern for transboundary pollution, a desire to provide humanitarian aid, and an interest in promoting exports. Yet in many cases, attempts at cooperation have conflicted with individual donor agendas. For example, as Connolly and I have noted, bilateral aid "features some of the most blatant examples of 'solutions chasing problems,'" when donors seek to finance activities that benefit their own companies and consultants or address environmental problems affecting their own borders.[34]

The Environment for Europe process is a loose set of institutional initiatives and regional meetings of environmental ministers, donor officials, and other interested actors from East and West that has become a broad political framework for cooperation on environmental protection in Europe.[35] Its strength and contribution stem from its role in creating an international policy network of government officials, donors, and NGOs that has increased the linkages and interactions among donors and recipients, while diffusing a set of ideas about how to set environmental priorities for the region.[36] It has encouraged convergence around specific policy processes, such as the use of environmental action plans, environmental impact assessments, and national environmental funds. At the same time, the ideas and processes espoused by this policy network are unevenly implemented within donor and recipient institutions, given the existence of divergent interests and weak institutional capacity, particularly within some recipient institutions. Donor institutions may work more closely together and increase their opportunities for cooperation, but the Environment for Europe process has not led them to dramatically change their aid strategies.

The World Bank and the EBRD have been involved with the Environment for Europe process directly in important ways, while the EIB has been a marginal player. The World Bank was the lead actor in organizing and writing the Environmental Action Programme for Central and Eastern Europe (EAP), which was one of the most important and visible outcomes of the Environment for Europe process.[37] It contained a set of guidelines, principles, and priority areas for environmental management in the region, and its goal was to be considered as a process and guide for policy development among Eastern and Western countries and multilateral institutions. The document was endorsed by the second meeting of Eastern and Western environment ministers and officials from international organizations at the 1993 Environment for Europe conference in Lucerne, Switzerland. Its main ideas are discussed below in the section on the World Bank.

In addition, the World Bank and the EBRD are involved in one of the two micro-institutions created by the Environment for Europe process to facilitate the implementation of the EAP: the Project Preparation Committee (PPC).[38] The PPC is a network of bilateral donors and the MDBs that acts as a matchmaker to bring together bilateral and multilateral assistance for projects that support the goals of the EAP. Its staff consists of a handful of officers at the World Bank and EBRD, and a small secretariat at the latter.

The PPC has helped to catalyze environmental assistance to CEE by promoting projects that blend (MDB) loan and (bilateral) grant aid in ways that are attractive to both these institutions and recipients. Recipients are more likely to agree to an MDB loan if it is sweetened by a grant component. Likewise, MDBs find they can more easily sell loan-based projects that include grant money. Bilateral donors, in turn, are pleased to link their activities with larger investment projects and can more easily counter criticism that they have funded studies that have led nowhere. At the same time, however, the PPC has changed donor priorities only at the margin. Most PPC projects consist of bilateral funding being added to MDB projects that were already in the banks' project pipeline. As a result, the PPC tends to fund projects that would have existed without it, although it may speed up the identification, evaluation, and implementation of specific projects.

By the late 1990s, the policy network created by the Environment for Europe process was being overshadowed by the EU accession process, which has given CEE applicant countries a powerful set of incentives to undertake extensive infrastructural investment and legislative reform. The negotiation process between candidate states and the EU includes efforts to extend the transition period for transposing and implementing the environmental *acquis*. In many ways, the accession process complements the Environment for Europe process and donors such as the MDBs are involved in advising and providing specific pots of money to assist in meeting the environmental *acquis*. Yet, there are also areas where the two clash, such as cases where the EU is calling for expensive higher standards where the Environment for Europe process is stressing cost effectiveness. Supporting CEE countries' arguments for longer transition periods is the fact that to date 40 percent the noncompliance cases brought by EU member states to the EU Commission have to do with environmental law.[39]

MDB Activities in CEE

Table 5.1 summarizes the three banks' lending commitments to CEE during 1990 to 1999, and reveals the basic details of their activities. Commitments reflect loans that have been approved by an MDB's board and signed by both the bank and the loan recipient. While commitment data are among the key figures in the banks' annual reports, they are mainly useful as a reflection of a bank's *intentions,* because commitment data alone do not reflect actual disbursements or loan cancellations.[40]

The table shows that the World Bank had the highest overall level of commitments, standing at $15 billion in IBRD and IDA funding, compared with around 8 billion euro ($10 billion) for the EBRD and 11 billion euro ($13 billion) for the EIB. Poland, Romania, and Hungary were the top three recipients at all three banks. The data also show that the World Bank was the leading lender in the majority of countries in the region, while the EBRD had the highest commitment levels in Estonia and Lithuania and was the only one with a significant portfolio of region-wide projects, which include a variety of multiproject facilities. The EIB, in turn, had the highest commitment levels in the Czech Republic, Slovakia, and Slovenia.

Data on the number of projects each bank is undertaking in a given country may not have significant meaning. It is difficult to be precise about the number of projects each bank is committed to undertaking, owing to the existence of multiproject facilities (at the EBRD) and projects that on-lend money to intermediaries that in turn lend to individual projects (at all three banks). And although the EBRD appears to have a significantly larger number of projects in most countries of the region than the other two banks, many of these are very small equity investments, capital increases to private banks, or other financial exercises not really comparable to large project loans dominating the portfolios of the other two.

The World Bank

General Approach

The World Bank's approach to CEE since 1990 reflects its broader mission of nesting lending and other policy activities into broad policy dialogues with recipient governments. Its lending program includes a substantial amount of economic and sector work, such as country economic memoranda, trade analyses, financial sector assessments, and social sector analysis, as well as energy and environmental studies for various countries in the region. As such, it is seen as a largely top-down strategy, with lending going to governments in the form of project lending or policy-based lending that is conditioned on macroeconomic reform.[41]

Many of the Bank's earliest loans to most countries in the region consisted of balance-of-payments support through large structural adjustment loans (SALs), aimed at addressing hyperinflation, balance-of-payments deficits, and other basic issues involved in the initial steps of transforming centrally planned economies into market-based economies. Other early loans focused on sectoral reform for the financial, agricultural, and energy sectors.

For countries in the earlier stages of economic restructuring, the Bank has supported liberalization, privatization, and public enterprise restructuring. In these countries, the Bank continues to support major fiscal adjustments through structural adjustment lending, and loans that help to develop basic social services and safety nets. For countries further along in the transition process that now attract significant private capital flows,

Table 5.1
World Bank, EBRD, EIB Commitments to CEE Countries 1990–1999[1]

	World Bank (IBRD, IDA, Trust Fund)		EBRD (loans, shares)		EIB	
	Total Commitment (millions/Dollars)	no. of projects[2]	Total Commitments (millions/ECU/Euro)[3]	no. of projects	Total Commitments (millions/ECU/Euro)	no. of projects
Albania	483 IDA only	33	109	11	68	3
Bosnia and Herzegovina	548 Trust Fund and IDA	25	75	9	0	0
Bulgaria	1210	18	324	24	699	15
Croatia	734	14	569	25	0	0
Czech Republic	626[4]	2	819	27	2072	20
Estonia	126	7	324	33	123	10
Hungary	2324	20	1368	54	1452	26
Latvia	315	14	271	19	193	11
Lithuania	293	12	244	17	232	14
FYR Macedonia	499 (294 is IDA)	17	222	11	130	0
Poland	4969	30	1646	76	2823	31
Romania	2971	25	1413	46	1448	24
Slovakia	135[5]	2	455	22	936	18
Slovenia	168	4	409	20	735	11

| TOTAL: | $15.4 billion | 8.3 billion | $9.9[6] | 10.9 billion | $13.0 |
| | $14.1 billion IBRD only | euro | bln | euro | bln |

Sources: World Bank, EBRD, EIB Annual Reports, 1990–1999; author correspondence with Josue Tanaka, EBRD; Yvonne Berghorst, EIB.

1. As noted in the text, commitments alone are an imprecise source of data, because many projects are canceled or reduced in size, which is not reflected in the data on commitments. It is also important to take the following into account when assessing these comparisons: (1) the World Bank's fiscal year runs from July 1 to June 30, whereas the other two banks use the calendar year; (2) because the EBRD approves framework agreements out of which smaller projects are funded, the number of projects counts operations as fractional numbers when multiple subloans are grouped under a single framework agreement. The EIB made its first loans to the region in 1989, and the EBRD in 1991. World Bank and EIB lending to CEE in 1990 consisted of loans to Hungary and Poland.

2. These figures for all three banks are also not precise. In some cases one bank document will list as one project what another document shows to be a group of subprojects.

3. The EIB and EBRD switched their reporting currency from ECU to euro on January 1, 1999, when the euro was adopted as the legal currency of 11 EU member states.

4. There were no World Bank loans to the Czech Republic after 1992.

5. There were no World Bank loans to Slovakia after 1992.

6. The conversion rate used for the EBRD and EIB is the average of the dollar to ECU/euro on December 31 from 1990–1999, or 1 Euro = $1.19. Individual loans in specific currencies are translated into an ECU/euro rate either at loan approval, or at the end of the quarter or year in which the loan was approved. Thus the actual ECU/euro rates of each project may fluctuate daily.

the Bank has adjusted its strategy to focus on support for eventual EU membership, which includes helping countries adapt to EU standards in a number of areas, including agriculture, industry, the environment, and social sectors.[42] For this group of recipients, as demand for World Bank loans declines, the Bank emphasizes its policy advice and guarantees.

The World Bank's Portfolio of Activities
Table 5.2 shows the sectoral breakdown of the World Bank's commitments to CEE in the period 1990–1999. It is important to note that this table and comparable tables for the other two banks incorporate projects with environmental objectives into broader sectoral categories where possible (i.e., putting wastewater treatment projects in the "water sector" or energy conservation projects into "energy sector"), since the banks do not have a consistent, comparable way of defining what is or is not an "environmental" project. Therefore, the "environment" category in table 5.2 only includes those World Bank projects that cannot easily be encompassed in a broader category, such as the Bank's forestry management and environmental capacity building projects. Environment projects from different sectors are placed together in tables 5.3, 5.6, and 5.8.

Over 20 percent of the Bank's loan commitments are for multisectoral and related lending, of which the vast majority are SALs and loans for basic economic restructuring and reform.[43] Loans supporting financial and industrial reform account for an additional 17 percent of the Bank's portfolio. These include a series of "enterprise and financial sector adjustment" loans for almost all countries in the region to support government programs on enterprise and banking reform. Energy and transport are major areas of lending, which is also the case with the other two MDBs. In the energy sector, the bulk of the Bank's activity is focused on power generation. The Bank's character as the least banklike MDB is also reflected in loans it makes for social sector, health care and population issues, and education, which distinguish it from the other two.

Approach to Addressing Environmental Issues
The Bank's strategy for addressing environmental issues in the region has two main components—its work in policy advice and reform, and its work in project lending. Policy work focusing directly on environmental issues undertaken before the Bank's mid-1997 reorganization tended to

Table 5.2
World Bank CEE commitments by sector, 1990–1999, in million \$[1]

Sector		Sector Total	% of portfolio
Multisector and Misc.		3195	20.4
Finance and Industry		2739.3	17.4
Energy:		2112	13.5
Power Generation	1525		
Oil and Gas	186		
Heat Restructuring/Conservation (District Heating)	101		
Mining	300		
Transport:		1822	11.6
Roads/Highways	901		
Railroads	316		
General Transport Sector	483		
Other (includes port rehabilitation, urban transport)	122		
Agriculture		1526.3	9.7
Public Sector Management		1362.3	8.7
Health/Population		538	3.4
Education		573.5	3.6
Urban Development		557.7	3.6
Telecommunications		465	3.0
Social Sector		310.3	2.0
Environment		305.2	1.9
Water Supply and Sanitation		195	1.2
TOTAL		15.7 billion[2]	100

Sources: World Bank Annual Reports, 1990–1999.
1. This table classifies loans based on their major purpose, although many loans are involved in numerous subsectors. Categorization by sector is also an imperfect art, which allows for a great deal of discretion. A number of financial sector loans, for example, reflect financing for agricultural reform, but are not included in the agriculture category, while some urban development loans support transport or water supply and sanitation, but are not included in separate categories. The data here includes IBRD, IDA resources, as well as Trust Fund financing for Bosnia and Herzegovina.
2. Discrepancy with table 5.1 due to rounding.

reflect the activities of staff in the central environmental department and the regional technical department. Project work emphasizing environmental goals, in turn, has also reflected the involvement of these staff, in addition to individual task managers who specialized in the development of projects with significant environmental goals.

In terms of its policy work, some of the Bank's more traditional economic restructuring policy advice has addressed environmental issues where it coincided with economic policy changes. An example would be the Bank's efforts, through its SALs, to persuade countries in the region to remove energy subsidies, in order to encourage a more sustainable use of resources. In developing its strategies for countries in the ECA region, the Bank included environmental specialists in the multidisciplinary teams sent to the region on mission.

More directly related to environmental policy reform are the Bank's efforts to help countries formulate environmental policy reform strategies and assist in relevant capacity-building exercises at the domestic level. Much of this work has been undertaken through the Bank's assistance in the development of national environmental action plans (NEAPs) and through its leadership in the development of the EAP.

The EAP argued that a mismatch between the high costs of environmental restructuring and the limits of Western aid requires donors and recipients to identify the highest priority environmental problems and to look for cost-effective ways of solving them. In other words, the needs of environmental aid and reform must be balanced against the costs of different reform options. Arguing that the level of damage to human health should be the main criterion driving the prioritization of environmental problems, it listed the region's major environmental priorities as: airborne dust caused by coal-burning; lead found in air and soil, caused by transport and smelting; and sulfur dioxide and other gases, emitted primarily by the burning of high-sulfur coal or fuel oil. The EAP emphasized that air pollution was probably the greatest short-to-medium term environmental problem in terms of its impact on human health, but also recommended addressing nitrates found in water, and drinking water contaminated by poor disposal of hazardous and nuclear waste. The problems identified as highest priority issues were primarily local, not transboundary.

The EAP argued that policy prioritization should be based on four criteria: (1) policymakers should support "win-win" economic reform policies that also have environmental benefits, such as liberalizing energy prices; (2) environmental policies should be targeted and should establish a framework of institutions and incentives to discourage emissions of pollutants and biodiversity loss in a cost-effective manner; (3) environmental spending should go to projects with the highest benefit-cost ratios; and (4) "modest" expenditures should be set aside for programs that have a long lead time from start to finish but still have high benefit-cost ratios.[44] The underlying message of the document was that problem prioritization, policy reform, institutional strengthening, capacity building, and investment were the areas both recipients and donors should emphasize. The key to environmental improvement in the region was thus not donor financing, but recipient policy reform.

The EAP disappointed governments in the region, who had hoped the effort would bring about additional donor funding targeted toward specific projects. While implementation of many of the EAP's strategies remains mixed, as discussed above, it is nonetheless important as a systemic attempt to set regional environmental priorities and to put forward a set of policy processes and specific policy solutions to address them.

In addition to its policy work on the EAP and on NEAPs, the Bank has been actively involved—sometimes playing a leading role—in a number of other multicountry or multi-actor regional initiatives involving transboundary water issues. These include action programs for the Danube River, the Baltic Sea, and the Mediterranean, as well as the Aral Sea, the Black Sea, and the Caspian Sea. These often involve the development of action plans, which lead to the financing of discrete projects by the donor community as well as recipient countries.

In terms of project lending, table 5.3 lists loans made by the Bank for projects with primary environmental goals or significant environmental objectives. In addition, the Bank has been responsible for carrying out an additional $88.9 million in GEF grant projects, which are also listed separately in table 5.4.[45] The 20 projects in table 5.3 total $1.3 billion, or just over 8 percent of the Bank's total commitments to the region. Out of this amount, the lion's share—eight projects totaling $959 million—was devoted to energy-related projects, which emphasized energy efficiency and the reduction of air pollution. Four loans, totaling $19.2

Table 5.3
World Bank (IBRD and IDA) projects with primary environmental goals or significant environmental components in CEE: 1990–1999

Year	Country	Project Title	Sector	Loan/ Credit $m	Co-Financed with other MDBs	Environmental Objectives
1999	Slovenia	Slovenia-Land Cadastre Project	Env.	15		Improve real property registration system, contribute to development of efficient real estate markets. Support of better government land use planning fosters more prudent use of land, natural resources. Supports improved information for environmental use, including orthophoto maps, aerial photography.
1998	Bulgaria	Environmental Remediation Pilot Project	Env.	16		To reduce environmental problems at newly privatized copper smelter. Addresses "leaking, arsenic-laden lagoon" threatening drinking water reservoir. Also addresses solid waste disposal.
1998	Croatia	Municipal Environmental Infrastructure Project	Water Supply/ Sanitation	36.3		Reduces discharges of wastewater pollution into Kastela and Trogir Bays, improves safety, delivery of drinking water.
1997	Latvia	Municipal Solid Waste Management Project	Env.	7.9	(GEF)	Demonstrates self-sustaining, modern management of municipal solid waste through maximum collection, utilization of landfill gas in city/district of Liepaja. Sets up unit to sepa-

					rate recyclables, segregate hazardous from municipal waste, collects methane gas and generates electricity from waste by-products. Reduces greenhouse gas emissions.
1997	Lithuania	Energy Efficiency/Housing Pilot Project	Urban Dev.	10	Improves energy efficiency in residential and public buildings and supports privatization of the housing sector.
1996	Lithuania	Klaipeda Geothermal Demonstration	Energy	6	Demonstrates the feasibility of using low-temperature geothermal water as a renew-able indigenous energy resource for district heating.
1996	Slovenia	Environment Project	Energy	23.9	Mitigation of air pollution and institutional strengthening. Will expand existing and build new district gas-heating networks.
1996	Croatia	Coastal Forest Reconstruction and Protection	Forest	42	Restore, protect coastal forests destroyed by war.
1996	Albania	IDA Forestry Project	Forest	8	Restore and promote sustainable use of degraded state-owned forest and pasture areas.
1995	Lithuania	Siauliai Environment	Water	6.2	Reduce pollution from upper Lielupe river basin, a major source of water pollution into the Gulf of Riga portion of the Baltic Sea.
1995	Estonia	Env. I—Haapsalu and Matsalu Bays	Water	2	Environmental management for coastal zone areas in western Estonia, which include re-habilitating wastewater and water systems, institutional strengthening.

Table 5.3
(continued)

Year	Country	Project Title	Sector	Loan/ Credit $m	Co-Financed with other MDBs	Environmental Objectives
1995	Latvia	Liepaja Env. Project	Water	4		Environmentally sustainable management of coastal zone, reduce wastewater discharge, and enhance water quality.
1995	Lithuania	Klaipeda Env. Project	Water	7		Environmentally sustainable management of coastal zone, reduce wastewater discharge, and enhance water quality.
1995	Poland	Katowice Heat Supply	Energy	45		Improve energy efficiency of district heating systems.
1994	Estonia	District Heating Rehab. Project	Energy	38.4	EIB	Improve energy efficiency of district heating systems.
1994	Poland	Forest Development	Forest	146	EIB	Promote sustainable forest management practices, soil erosion prevention, increase vitality of forest stands, modernize management operations, equipment.
1992	Czech Republic	Power and Environmental Improvement	Energy	246		Reduce air pollution at Prunerov power plant through desulphurization equipment installation.
1991	Poland	Heat Supply Restructuring and Conservation	Energy	340	EBRD	Restructure energy sector, privatize and commercialize, promote energy conservation, reduction of air pollution, technical assistance and training.

1990	Poland	Energy Resource Development	Energy	250	EIB	Increase gas production and energy conservation, reduce air pollution, strengthen regulatory, institutional framework.
1990	Poland	Environmental Management Project	Misc. (capacity building)	18		Strengthen national, local environmental management institutions. Develop air quality management strategy for Katowice-Krakow region, integrated framework for river basin management in upper Vistula River, build local capacity for better donor coordination.
TOTAL				1,267.7		

Sources: Annual Reports; World Bank, "The World Bank and the Environment in Central and Eastern Europe: 1990–95" (Washington, D.C.: World Bank, 1995); World Bank, "Environment Matters: Annual Review" (Washington, D.C.: World Bank, Fall 1997); World Bank, "The World Bank and the Environment in Central and Eastern Europe and the Commonwealth of Independent States: 1995–1998" (Washington, D.C.: World Bank, 1998).

Table 5.4
World Bank GEF projects in CEE: 1990–1999

Year	Country	Project Title	Sector	Amount $m	Project Description
1999	Romania	Biodiversity Conservation Mgmt.	Biodiversity	5.5	Develop, implement management plans for three top sites identified by Romanian Biodiversity Steering Committee.
1998	Latvia	Solid Waste Mgmt. and Landfill Gas Recovery	Short-term measure	5.1	Linked to IBRD loan of $8.0. Project remediates landfill, installs technology to enhance biodegradable waste degradation and uses collected landfill gas for power generation.
1998	Czech Republic	Kyjov Waste Heat Utilization	Short-term measure	5.8	Use gas-fired, combined cycle cogeneration system at a bottle manufacturing factory. Part of waste heat will be used to produce electricity for factory, and part used for local district heating system.
1997	Poland	Ozone Depleting Substances Phase Out	Ozone	6.2	Phase out about 55% of country's 1994 weighted ODS consumption through projects in refrigeration, foam-blowing, and medical aerosol industries.
1996	Lithuania	Klaipeda Geothermal Demonstration	Global Warming	6.9	Help finance construction of demonstration geothermal plant to provide hot water for district heating system.
1995	Hungary	Ozone Depleting Substances Phase Out	Ozone	6.9	Phase out 50% of country's 1993 annual, ODS consumption through subprojects in solvents, foam, aerosol, halon and refrigeration sectors, and through recovery/recycling efforts.
1995	Slovenia	Ozone Depleting Substances Phase Out	Ozone	6.2	Phase out around 36% of country's 1993, annual weighted ODS potential through subprojects in refrigeration, foams, aerosol, solvents sectors.

Year	Country	Project	Focus	Amount	Description
1995	Bulgaria	Ozone Depleting Substances Phase Out	Ozone	10.5	Phase out 65% of country's 1993 annual, weighted ODS consumption through projects in refrigeration, foam blowing and solvents sectors.
1994	Poland	Coal-to-Gas Project	Global Warming	25	Extend coal-to-gas conversions to medium-size boilers, to demonstrate interfuel substitution, technological innovation to reduce CO_2 emission.
1994	Romania	Danube Delta Bio-diversity	Biodiversity	4.5	Protect ecosystem and contribute to biodiversity conservation.
1993	Czech Republic	Biodiversity Protection	Biodiversity	2	Protects biodiversity in three ecosystems, supports activity of three transnational biodiversity protection networks.
1994	Czech Republic (?)	Ozone Depleting Substances Phase Out	Ozone	2.3	Eliminate production of CFCs, establish national refrigerant recovery/reclamation/recycling program.
1993	Slovakia	Biodiversity Protection	Biodiversity	2	Develop management techniques, conservation program for biodiversity protection in Morava floodplain, Tatras forests and Eastern Carpathians.
TOTAL				88.9	

Sources: World Bank, "The World Bank and the Environment in Central and Eastern Europe: 1990–95," (Washington, D.C.: World Bank, 1995); World Bank, "Environment Matters: Annual Review," (Washington, D.C.: World Bank, Fall 1997); The World Bank and the Environment in Central and Eastern Europe and the Commonwealth of Independent States: 1995–1998 (Washington, D.C.: World Bank, 1998); World Bank–GEF Full Project Portfolio, http://www.-esd.worldbank.org/gef/fullProjects.cfm.

million, consist of projects in the water sector in the three Baltic states, which focus on reducing water pollution in the Baltic Sea by rehabilitating water and waste-water systems and also contain components for the environmentally sustainable management of coastal zone areas. The Bank is also involved in sustainable forestry loans in Albania, Croatia, and Poland. In addition, it funded an environmental management project in Poland aimed at strengthening domestic environmental institutions. This was the Bank's first project in CEE following the regime changes. In recent years, it has also branched out into solid waste and land management projects.

Given the fact that nearly all MDB projects have multiple objectives, what distinguishes the projects in table 5.3 from others the Bank undertakes? Investments to improve infrastructure are important to all countries in the region, and MDB "environmental" projects tend to revolve around water and energy infrastructure rehabilitation. In this sense, virtually all of the MDB environmental projects might be considered to be "business-as-usual" investments. Yet, a closer look at the projects in table 5.3 (as well as the comparable tables for the other two MDBs) reveals qualitative difference in project objectives between these and other projects in the banks' broader portfolio. In terms of the energy projects on this list, as noted above, some important indicators of an emphasis on environmental objectives in energy lending include the extent to which a project addresses supply-side efficiency or demand side management and conservation issues. This moves projects beyond more traditional MDB loans for building or rehabilitating power plants, or expanding transmission and distribution systems.[46]

The Bank's more "green" energy lending in CEE consists of both sector loans and project loans. The sector loans, such as the Poland Energy Resource Development, by definition involve broader sectoral restructuring and policy reform. The main goals of the Polish loan were to increase the country's output of natural gas through investments to improve gas recovery from existing gas fields, as well as to evaluate other potential gas fields; to encourage greater domestic use of gas versus hard coal, which provides the vast majority of Poland's energy supply; and to encourage energy conservation through the support of energy price reform. The project's designers expected that gas produced under the project would have an impact on reducing the amount of hard coal used for household heating, which in turn would reduce emissions of ash, sulfur dioxide, carbon dioxide, and nitrogen oxide.[47]

The other energy projects on this list also include specific environmental goals or components. The Czech "Power and Environmental Improvement" project, produced out of the Bank's "Energy and Environment" division in the country department addressing the Czech and Slovak republics, was designed to improve the efficiency of power plants in northern Bohemia in order to reduce energy costs and air pollution.[48] This $246 million loan to the state-owned Czech Power Enterprise (Ceske Energeticke Zavody, or CEZ) was one of the largest project loans designed by the Bank in the region (in contrast with SAL or sector loans).[49] The largest investment components of the project (almost 70 percent) were related to pollution control, including the installation of FGD equipment, dust control equipment, as well as other operational improvements to reduce lignite consumption.[50]

Several of the Bank's energy projects involved improving the energy conservation and efficiency of district heating systems, which are major sources of local air pollution in the winter heating season, since they typically rely on often inefficient boilers that burn fuels such as heavy fuel oil, coal, natural gas, shale oil, wood, or peat.[51] These projects tend to be part of broader Bank involvement in energy sector reform and also support the recommendations of the EAP. The Estonia project, for example, was set up to improve the efficiency of district heating in several cities to help the country reduce fuel costs and to improve the environmental conditions in affected areas. More specifically, it involved a small boiler conversion and replacement program in a number of municipalities across the country, to help them switch to local fuels in order to reduce heat production costs; efficiency improvements in Tallinn, Tartu, and Pärnu that included equipment and boiler rehabilitation, as well as the installation of other equipment to improve efficiency (such as meters, variable speed pumps, and new substations); and institutional support for relevant Estonian agencies, including training, equipment, and software for project management.[52]

The Slovenia Environment Project also focused in part on making heating systems more efficient, but with an even "greener" slant. One of the project's two main goals was to reduce air pollution by providing loans to around 6,500 households and 65 boilerhouse operators to cover 80 percent of the cost of converting their fuel or heating systems from polluting fuels (such as coal, wood, other solid fuels) to cleaner fuels (natural gas, solar energy). The Bank lent money to Slovenia's EcoFund for onlending to residents of six cities, following a pilot phase in Maribor, Slovenia's second largest city, which decided in 1992 to begin converting all of its heating systems to cleaner fuels. The project's second objective was to assist the government in developing a national geographical information center within the environment ministry, which, in turn, would manage a geographical information system (GIS).[53] The task manager for this project was one of the "green task managers" working on energy and environmental projects in the region.[54]

Other "green" energy projects include district heating projects in Poland and Latvia, as well as one in Lithuania that focuses on using low-temperature geothermal water as a renewable energy source. The geothermal project was seen as a "demonstration" project to determine the feasibility of providing renewable energy to help reduce Klaipeda's district heating utility's use of oil to heat its water.[55] These projects ultimately emphasized environmental issues more than the broader portfolio of World Bank energy projects in Central and Eastern Europe. The broader energy portfolio includes a project in Poland to modernize high voltage electricity transmission lines and substations, a Bulgarian project to reduce the operating costs and improve the efficiency of the national electricity company, and a project in Lithuania to rehabilitate two thermal power plants to improve the safety and flexibility of the transmission system.

Turning to the water projects on the list, the four Baltic projects can be distinguished from more typical MDB projects in water sanitation and supply by the fact that they were all designed to address environmental management of "hot spots" identified by the 1992 Joint Comprehensive Environmental Action Programme (JCP) for the Baltic Sea as priorities for environmental clean-up.[56] All four projects contained water sanitation and supply components, largely funded by Bank loans, as well as specific "environmental management components" (EMCs), funded mostly by bilateral grant money. These EMCs focused on issues related specifically to the environment, such as coastal zone management. The goals of all four projects were to reduce the discharge of wastewater into the Baltic Sea through the improvement of wastewater and water supply services, while supporting the development of management plans for the sustainable development of nearby coastal areas and wetlands. The designer of three of the four Baltic projects was a senior environmental specialist at the Bank, who was also the World Bank's coordinator for the regional Baltic Sea Environmental Programme. These projects were managed by a small Bank team set up to work on Baltic water projects.[57]

In comparison to these projects, other World Bank water projects in Central and Eastern Europe do not have the same emphasis on environmental components, although they may have environmental benefits. As an example, the $98 million water companies restructuring and modernization project in Bulgaria was designed primarily to help a set of municipal water companies gain autonomy from the central government, in the process becoming more accountable to local authorities, more commercially oriented, and more financially sound. The project supported the Bank's goal of strengthening Bulgaria's infrastructure. It also had environmental benefits, since if implemented properly, it would encourage a more efficient use of water and a reduction of pollution from wastewater. However, these components were not major factors driving the project forward. Unlike the Baltic projects, this project was developed within the Bank's infrastructure division within the country department that addressed Bulgaria.

In addition to the individual energy and water-related projects, the Bank also funded an innovative environmental management project in Poland, designed to strengthen the capacity of Poland's environmental institutions in the analysis and design of policy, regulations, and investment actions to improve environmental quality, as well as fund-specific investments in air quality management and river basin management in priority areas. This 1990 project sought to help the country's environmental ministry implement a management information system and improve administrative procedure for budgeting and work programs. Other project subcomponents focused on improving government policy and information on a variety of environmental issues, including issues such as environmental health, municipal solid and hazardous waste, air quality, and water resource management. The project created a project implementation unit within the environment ministry, to coordinate the project as well as environmental assistance from other donor institutions. This project was also developed by staff from the Bank's central and regional environmental departments.[58]

To summarize, the Bank's "environmental" portfolio in Central and Eastern Europe emphasized—at least on paper—environmental goals or issues more strongly than other projects the Bank undertakes in the energy and water sectors that may also have positive environmental benefits. These particular projects are usually developed by the Bank's green bankers, people whose job focuses on identifying environmental investments or policy exercises.

Role of Recipient Demand

Although the World Bank is the least demand-driven of the three MDBs, recipient demand still affects its ability to sell investment projects in the region. The further along the transition process a recipient country is, the more alternatives the government will have in its quest for development financing, and the less pressure it has to accept funding with policy conditionality attached to it. As a result, in Central and Eastern Europe, the World Bank has been more successful in selling environmental conditionality to countries earlier on in the transition process, or for projects where recipients are specifically interested in World Bank expertise. The Czech Republic, for example, has only three World Bank projects and has not borrowed from the Bank since 1993.[59] The World Bank's attempt to develop an environmental capacity building loan there, similar to the Polish loan, was unsuccessful.

In the more developed countries such as Poland and Hungary, there are also examples of environmental projects developed by the Bank that the countries chose to finance through other institutions, or projects the Bank turned down on environmental grounds that other MDBs snapped up. One result of this is that the Bank has had to adjust its strategies for working in these countries, and by the mid-1990s it began to emphasize policy advice for the wealthier countries seeking to join the European Union. The World Bank's darker green environmental objectives can become a disadvantage to its lending programs in markets that have become more competitive.

At the same time, where recipients are receptive to the Bank's work, the Bank often has much power to influence the specific shape that projects take. Slovenia, for example, wanted to buy FGD equipment to reduce air pollution from its power plants. Bank officials argued that this was not cost effective vis-à-vis other options and proposed a coal-to-gas conversion project that the Slovenians agreed to, which evolved into the Slovenia environment project discussed above. "Countries often come in with ill-conceived ideas," noted one senior World Bank environmental economist, who added that the Bank "gently persuades" them to adopt more "sensible" policies or projects.[60] The Bank also turns down projects of interest to recipients that do not meet the Bank's own interests or standards (whether economic, environmental, or other), such as dam projects in Slovenia and Croatia, an aluminum smelter in Slovakia, and a project to expand Budapest's metro. The EBRD, in turn, took on the Slovak project rejected by the World Bank, and the EIB funded the Budapest metro project.

The four Baltic water projects provide an instructive illustration of how the Bank can influence what recipients agree to do.[61] In these cases, the Baltic municipalities were mainly interested in local issues, such as improving poor wastewater treatment and/or the drinking water supply, as well as attracting more tourists by cleaning up nearby coasts. The broader issue of cleaning up the Baltic Sea as a whole was less of a pressing concern to the municipalities than to their central governments, the Bank, and other (particularly Nordic) donors. The environmental management components of the projects were generally not a priority to these local governments, and as a result, the Bank "sold" them to recipients not as loans that would have to be repaid, but as grant components supplied by bilateral donors. There was therefore no cost to recipients for agreeing to projects that included EMCs. These EMCs were also very small. For example, for the $23 million Siauliai project, the EMC totaled $1.4 million, and for the $21 million Liepaja project, it totaled $1.5 million.

At the same time, the water projects designed by the Bank almost always went beyond what the municipalities initially had in mind, as the World Bank tacked on additional components to pursue a broader watershed approach. The Bank had particular leverage to do so with these projects, because most could not find alternative sources of financing at the time they were designed, often even from other MDBs. The World Bank also brought bilateral donors into financing parts of the wastewater and drinking water components of the loans as well.

In Haapsalu, Estonia, for example, the municipality hoped to complete construction on a biological wastewater treatment plant to replace the existing mechanical treatment plant. Construction on the plant began before Estonia achieved independence in 1991, and after that, the Estonians did not want to complete construction on the Soviet-designed plant with Russian technology. Estonia was unable to secure financing for the project from the Swedish aid agency or the EBRD. The World Bank, however, expressed an interest in financing the project if it included an EMC that focused on nearby Matsalu Bay, which itself was on the HELCOM list of priority hot spots as an important wetland and nature reserve.[62]

The municipality of Siauliai, Lithuania, in turn, was keen on finishing its partially constructed wastewater treatment plant and addressing the problem it faced by its inadequate capacity to handle sludge.[63] Siauliai is the fourth largest city in Lithuania, one of its main industrial centers, and home—before Lithuania's independence—to the largest Soviet military airport in Eastern Europe. Like the other cities, it had a partly constructed Soviet-style plant that was meant to replace an older plant that no longer met the city's needs.[64] However, the design of the unfinished plant was too big for the city.

The World Bank project, instead of confining itself to the completion of the municipality's wastewater treatment plant, added a number of additional environmental components to the project. These included an EMC that would organize bilateral funding for training, technical assistance and other support for a Lielupe River Commission set up by Lithuania and Latvia in 1993 to cooperate on issues concerning management and clean up of the river.[65] This component would also provide technical support, equipment, and training for two Lithuanian regional environmental protection offices; assistance in the development of procedures to monitor discharge; a plan for sludge management; and a large study complemented by demonstration activities aimed at improving the practices of large local pig farms in managing agricultural run-off.[66] The Bank argued that better practices by local pig farmers would greatly enhance the project's benefits in terms of reducing pollution to the Baltic Sea.[67]

In sum, while recipient demand determines the countries or sectors in which the Bank can work, the Bank remains proactive and influential in pursuing particular project goals and designs, particularly in cases where alternative financing sources—with less conditionality attached—are not available.

On Implementation

Project implementation for any MDB rarely occurs as it is supposed to on paper. Delays can arise at any level, estimates and projections that underlie the project may prove to be overly optimistic or pessimistic, and unanticipated technical and political problems are not uncommon. Projects can stall or even fail at any MDB for myriad reasons, including changes in governments, weaker-than-expected domestic institutions, over-optimistic implementation schedules, poor project management, changing external economic conditions, or the absence of clearly specified requirements. The World Bank's Quality Assurance Group, for example, lists twelve different criteria for "projects at risk," which include slow disbursements, country risks, poor economic management, risky subsector, poor environmental or resettlement performance, poor compliance with legal covenants, project management problems, procurements problems, and poor financial performance.[68]

Project performance is also shaped by how strict the performance objectives are. Loosely defined projects can appear to be complying better than projects with more complex goals and conditionality, even though they might have less development impact. As Albert Hirschman noted over thirty years ago in studying a set of World Bank investments, "It quickly became apparent to me that all projects are problem-ridden; the only valid distinction appears to be between those that are more or less successful in overcoming their troubles and those that are not."[69] As noted above, the data necessary to evaluate each MDB project is not readily available, often even within the banks themselves.[70] This section draws on data that is available on World Bank projects in CEE, as well as interviews with NGOs, World Bank officials, and recipient country officials, to sketch out at least part of the implementation picture.

Of the three MDBs, the World Bank provides the most information on implementation, as discussed in chapter 4.[71] It provides cancellation data, although the data does not state the reasons for cancellation. This can make significant difference in the actual flows of funding going to individual countries. For example, almost $1 billion of loans to Poland's $5 billion portfolio has been canceled.[72] Although none of the projects in table 5.4 has been completely canceled, some have faced problems or partial cancellation. Among the projects discussed in the previous section, three have had portions canceled. These include the $98 million Bulgarian water companies restructuring project was reduced by $41 million; the Polish forestry project, reduced by $42 million; and the Polish Heat Supply Restructuring project, reduced by $75 million.

A 1997 in-house review of fourteen World Bank sectors by the Bank's Quality Assurance Group highlighted a handful of problematic projects in CEE. These included seven projects in "extended problem status" and projects scheduled to close by mid-1997. None of these were projects whose main objectives were environmental, or were problem projects because of environmental issues.[73] Certainly, the complexity of the transition process resulted in numerous implementation problems throughout the region. The 1998–2000 Hungary CAS, for example, noted that Hungary was placed on a list of countries with problem portfolios, with five projects totaling $442 million in trouble.[74] One of the lessons learned, it said, was the importance of sequencing reforms into "manageable steps," since an overly ambitious program does not leave time to build the needed political consensus. One example of poor sequencing was a project to strengthen the country's pension administration that began working on institutional strengthening before any pension reforms had been identified. Other problems included project approval occurring before basic requirements for implementation had been addressed, weak government commitment to specific project objectives, projects that were not focused, and poor project supervision.[75]

Sometimes World Bank evaluations highlight projects that are doing particularly well. The Hungary energy development project was praised with respect to a broader set of oil and gas projects for thorough project preparation, solid sector studies, well-defined objectives, and clear government commitment.[76] The small $18 million Polish environmental capacity-building project also received a "highly satisfactory" review in its implementation report. According to the report, the project "achieved its major objectives, exceeded its physical objectives and is likely to achieve sustainable development results without major shortcomings."[77]

A more coherent picture of the Bank's performance in one country emerges from an in-depth analysis by the Bank's Operations Evaluation Department 1997 review of the Bank's activities in Poland. According to the review, the Bank's strategy in Poland between 1986 and 1996 was "highly relevant," while its efficacy was "satisfactory" and its efficiency only "marginally satisfactory."[78] The report argued that in the early years of the Bank's activities in postcommunist Poland, its advice, intellectual assistance, and technical inputs were pivotal in assisting the government in the design of economic reform, while its project loans were often relatively less valuable. For example, much of the Bank's initial lending to Poland was made through financial intermediaries, which were supposed to on-lend the funds to the final beneficiaries. However, these Polish intermediaries were generally weak and lacked the institutional capacity to adequately perform their on-lending tasks.[79] In addition, the report argued that the Bank's Polish team faced pressure to lend during the first two years of lending, which meant that some projects were poorly prepared.

Implementation was often slow in these years, which also reflected the country's changing political and economic context. As the report noted for the 1992–1993 period: "It became apparent that some projects were too large and complicated, involving several ministries or agencies that did not always share the same objectives. The turnover of government personnel also revealed some gaps in the Bank's institution-building efforts. New officials had difficulty learning the Bank's language and procedures. Communications between Polish officials and Bank staff were not always smooth. Lack of continuity in Bank staff sent from headquarters also contributed to implementation delays."[80]

The result was that the Bank restructured its portfolio in Poland, dropping some unsuccessful activities. The Bank's strengths were in areas of institution-building, such as foreign debt management and the improvement of public sector management practices. It was less effective in its attempts to push the government to sustain the pace of privatization or bring down inflation. Its strategy would also shift increasingly toward policy advice and institution-building rather than resource transfers, since Poland's impending accession to the European Union would reduce the Bank's leverage in the country.

Turning to the Bank's activities in specific sectors, the report's findings were mixed. In the energy sector, it argued that both the government and Bank were "overoptimistic" about sectoral reform.[81] Tariff increases were slower than agreed on, and the passage of new energy legislation was also slower than expected, while the government was hesitant to open up areas of the energy sector (such as power generation) to private sector involvement.[82] In addition, many reform issues associated with this sector—such as de-monopolization and commercialization—were politically sensitive and faced strong resistance from those who would be hurt by the reforms (such as the coal sector's unionized labor force).

In terms of the Bank's environmental activities, the OED report praised the Bank's Polish team's work on institution building, which strengthened the Ministry of Environment's ability to coordinate and manage foreign aid. It also noted the Bank's unwillingness to fund investments in energy, mining, and other sectors unless they complied with environmental guidelines. However, the report also pointed out that the government's commitment to environmental management was the "driving force behind the establishment of funding mechanisms that led to a high level of investment in pollution abatement."[83] The Bank's role was clearly supportive, although its direct contribution to a decline in various measures of pollution is more difficult to assess.

Interviews with environmental groups and government officials in the region generally support the main themes of the Polish report, with respect to the Bank's work in the region as a whole—that is, the Bank's main contribution in providing intellectual and policy support in the area of government policy reform.[84] Portfolio performance tended to be more mixed in countries struggling to a greater degree with the transition process, such as Bulgaria and Albania, and individual projects in each country are also the source of criticism or praise. There are no environmentally disastrous World Bank projects in CEE, reminiscent of the sorts of projects in Asia and Latin America that have galvanized NGOs.

Interviews with ministry officials and NGOs in CEE revealed, however, that MDB projects in the implementation stage are often not closely monitored by government actors and NGOs. Government ministries in the countries do not have the staff or technical capacity to carefully monitor MDB projects. The major environmental groups, in turn, tend to focus on projects in the preparation stage, or early implementation stage, to try to influence project design and act as watchdogs against environmentally unsustainable projects. They also do not have the capacity to follow projects through implementation.

Responses to the projects in table 5.3 were mostly positive, but some project problems were also pointed out. Among the positive responses, Polish environmental officials credited the Environmental Management project with exposing the Polish environmental policy community to the World Bank's ways of thinking and working, and setting up a project implementation unit at the ministry that helped to implement donor projects more broadly, while training a group of policymakers. In terms of project problems, World Bank officials noted that the 1990 Polish Energy Resource Development project did not meet its goals, since it resulted in increased gas imports, rather than increased gas production. Conditionality related to the privatization of the Polish Oil and Gas Company was also not met, according to officials involved in project implementation.[85] The Bank's Klaipeda Geothermal Demonstration project, in turn, fell behind by more than a year, owing to the fact that once well-drilling began, the geothermal company financed by the Bank learned that the water temperature was not as hot as expected, which required it to drill an extra well.[86] Environmentalists also criticized the project for not including a component to reduce the high levels of energy wasted in Klaipeda's inefficient district heating system, to which the project was providing energy.[87] NGOs were also critical of the fungibility of the Bank's loan to Czech Power Enterprise (CEZ), arguing that it freed up CEZ to increase its own investment in nuclear projects that were neither part of the project itself, nor something the World Bank would fund directly. According to CEE Bankwatch Network, two of the five projects for which CEZ used World Bank loan resources were connected to the Temelin Nuclear Plant, a controversial, partly-built, Soviet-era nuclear plant that the Bank had previously advised the Czech government not to complete.[88]

The Bank's water projects have faced some problems common to other MDB water projects in the region, as well as some particular implementation problems. MDB water projects throughout the region have faced the difficulty of increasing tariffs to rates called for in project documents. Tariffs are the major source of revenue for the water utilities, and therefore are also the source of financial resources used to repay the MDB loans. In the Baltic states, for example, all of the MDB-financed water utilities faced sharper-than-expected declines in water demand, which in turn reduced expected revenues that could be collected from tariffs. Water utilities also struggled to cut costs to levels needed to meet the financial requirements of the projects. In addition, it was politically difficult for some municipal governments to agree to increase water tariffs as required under the loans. As a result, many of the MDB water projects did not perform as well as expected in financial terms.

One weakness of the World Bank's Baltic water projects in particular has been that the projects' small environmental management components, funded by bilateral donors, have had mostly minor, if any, tangible results. They produced some studies that may or may not be useful in future policy development, such as integrated coastal zone management plans in Lithuania and Latvia. The Bank canceled the EMC for Siauliai concerning the Lielupe River Commission because of insufficient local interest. At the same time, a few components of the EMCs have been viewed as successful. One is the pig farm study in Siauliai, which produced recommendations on practices that were adopted by Lithuania's pig breeders' association. Another is the EMC for the Liepaja project, which included the purchase of a reed harvester for Latvia's Lake Pape, a large coastal wetland that is part of the coastal wildlife habitat. The harvester allowed the local municipality to harvest reeds that clog up the lake and sell them in Denmark and Lithuania, where they are used for roof material.

While the Bank's projects in Latvia and Estonia performed largely as expected, the projects in Lithuania faced relatively more challenges. Some of the problems were outside of the Bank's control, such as a rapidly changing group of Lithuanian policymakers (including three different environmental ministers in the space of six years). The Klaipeda project was delayed as a result of slow procurement tendering after the Ministry of Environment insisted on a retendering in order to involve more Lithuanian firms. Shifting government requirements on waste effluents delayed the project for another twelve months. Finally, both sides realized that the Bank project overlooked the need to include a very expensive ($1 million), important piece of equipment—the transformer to connect the water utility to the power grid.[89]

Disagreements between the Bank and the Lithuanian environment ministry also delayed the water projects. One general debate was over whether or not the wastewater plants needed back-up systems (such as pumps, or piping), which were required under the old regime and by old Soviet standards, since equipment often broke down. Such additional equipment adds to project costs and is generally not required in Western-style plants.

In some cases, disagreements between the Bank and municipalities were about issues where the Bank's stance would reduce the possible environmental benefits of the project. For example, both Klaipeda and Siauliai municipalities were interested in wastewater plants that included processes for removing nitrogen and phosphorus, which the World Bank argued was too expensive to undertake as part of the projects. The Bank determines what it thinks the capacity of the municipality is to take on debt, which Bank officials said results in projects that do not contain "all the bells and whistles" the municipality might desire.[90]

In Siauliai, in particular, there were many disagreements between the Bank and the municipality, and the project eventually stalled. Some municipal officials were angry that the Bank designed the project with what they perceived to be unnecessary components, while it lacked some major municipal priorities, such as sludge management or nitrogen and phosphorus removal. The Bank's view was that such components were not least-cost and were more expensive than the municipality could afford at the time. The Bank also argued that after the project was signed, the municipality decided it wanted duplicate equipment that would raise project costs. The Bank believed this equipment was unnecessary and reflected old Soviet standards, and it refused to increase the project loan size.

These anecdotes illustrate the fact that even projects that look right on paper may encounter many kinds of political, economic, and technical problems in practice. The World Bank has had to push harder than the other two to sell particular environmental projects to recipients, but even after these projects are agreed, it is often difficult to know exactly what impact they have on environmental improvement. Many potentially important contributions, such as advice, technical assistance, and research, are the most difficult to measure.

EBRD

General Approach
The Bank's portfolio clearly reflects its mission of promoting private sector development in CEE, primarily by emphasizing privatization, financial sector restructuring, infrastructure development, and the promotion of foreign investment. As table 5.5 shows, more than a quarter of the Bank's commitments of loans and equity investments in CEE is targeted to financial institutions, largely for developing the institutions themselves and for providing funding to be on-lent for the development of private sector investments, mainly for small- and medium-sized enterprises (SMEs). Many of these loans actually benefit other categories in the table, such as credit lines for agribusiness, manufacturing and energy efficiency.

Table 5.5
ERBD CEE commitments by sector, 1991–1999, in millions euro

Sector		Sector Total	% of portfolio excluding regionals
Banks/Credit Lines for SMEs/Capital Funds		2508.3	28.1
Manufacturing/Joint Ventures/Industry		1788.6	20
Transport:		1347.7	15.1
Roads/Highways	710		
Railroads	352.2		
Airports/Aircraft	90.2		
Other (includes combination projects)	195.3		
Telecommunications		1107.6	12.4
Energy:		973.3	10.7
Power Generation/Distribution	578.9		
Oil and Gas	216.6		
ESCO/Energy Efficiency	177.8		
Water Sanitation and Supply		367.6	4.1
Agriculture/Food/Beverage		399.2	4.5
Real Estate/Construction		171.4	1.9
Hotels		86.3	1.0
Wholesale/Retail/Misc.		90.6	1.0
Insurance		9.5	.1
Misc. Environment		48.8	.5
TOTAL:		8.9 billion euro[1]	99.4[2]

Sources: EBRD information sheets, Annual Reports (1991–1999), EBRD web site (www.ebrd.com/opera/country/)
1. Discrepancy with table 5.1 due to rounding.
2. Discrepancy due to rounding.

An additional 20 percent of the EBRD's portfolio supports joint ventures and domestically owned businesses and includes projects such as the construction of a Coca-Cola bottling facility in Albania, the establishment of a wholesale market system in Bulgaria, and the modernization of a paper and pulp facility in Poland. Transport is the third largest area of Bank commitments, where the majority of the EBRD's projects have emphasized highway construction or rehabilitation, often designed as toll roads.

As the private sector evolves in CEE, an increasing percentage of the EBRD's portfolio is directed toward private sector actors and activities, often in areas not addressed by the other two banks—such as constructing breweries, modernizing cotton fabric mills, and renovating safety facilities at a car safety belt manufacturer. In Hungary and Poland, for example, around 90 percent of the Bank's commitments are made to private enterprises or state-owned enterprises in the process of being privatized. In countries that have made less progress in the transition process, such as Bulgaria, Romania, and Albania, the Bank's strategy focuses more on sovereign-guaranteed investments in infrastructure, financial sector restructuring, and privatization support.

As noted in the World Bank section, project commitments do not reflect cancellations or delays, so actual disbursements vary from the picture painted by commitments. Of the projects that have some type of environmental focus, discussed in greater depth in the following section, Bank staff have reported that only 13 million ECU has been disbursed of the 40 million ECU Bulgarian Maritza East Power Project, which was approved in 1992. No disbursements had been made by mid-1997 on Albania's Drin River Rehabilitation project, which was approved in 1994. Small amounts of the Latvian energy sector emergency loan were also canceled.

Approach to Addressing Environmental Issues

Most of the EBRD's environmental attention has been on project lending, although the Bank has undertaken a handful of policy-related exercises. For example, it sponsored a research project on the implications of the harmonization of CEE environmental standards with the EU's, as well as institutional changes needed to improve environmental monitoring and enforcement, in part to assist local and foreign investors in making investment decisions. The EBRD's role as secretariat of the Environment for Europe process's Project Preparation Committee (PPC) has also involved it in some research activities in areas such as green equity financing.

Turning to project lending, table 5.6 highlights EBRD projects with primary environmental goals or significant environmental objectives. It includes both investment projects in the region, as well as a number of regional multiproject credit facilities that in turn finance smaller projects both in CEE and the other former Soviet countries. The total of CEE projects and regional MPFs in table 5.6 is 979 euro, or around 12 percent of the Bank's total portfolio of projects and regional credit lines in CEE.[91] The two biggest categories of projects in this list are in the water and energy efficiency areas, accounting for 40 percent and 31 percent of the total, respectively. Generally, the projects on this list come out of the Bank's municipal and environmental infrastructure unit, which finances district heating projects and water and environment projects; and its energy efficiency unit, which finances the ESCO projects in the region. The Bank's energy sector team, in turn, tends to finance the larger power sector projects, such as those rehabilitating thermal power plants and transmission systems.

The EBRD's energy efficiency projects, designed by the Bank's energy efficiency unit, emphasize the development of ESCOs and the use of credit lines for energy efficiency. This differs from the World Bank's approach to energy efficiency, which has focused more on broader sectoral restructuring and policy reform. The EBRD has signed a number of multiproject facilities (MPFs) with Western companies to set up ESCO companies in CEE. Typically, these are loans to companies like France's Compagnie Générale de Chauffe or the United States' Honeywell, which have expertise in energy controls and related environmental services. These companies, in turn, finance individual ESCOs in CEE, in which the EBRD takes a minority share.[92] The ESCOs provide energy saving services to clients that may include hospitals, district heating companies, and other institutions.[93]

The Bank has also financed a handful of other credit facilities for energy efficiency and municipal projects, which has allowed it to finance projects that would be too small for a normal EBRD loan. One example is the 14.5m DM credit line the EBRD extended to Priemyselna Banka Kosice

Table 5.6
EBRD projects with primary environmental goals or significant environmental objectives in CEE: 1990–1999

Year	Country	Project Title	Sector	Loan (l)/ Shares (sh) millions euro	Co-Financed with other MDBs	Environmental Objectives
1999	Regional	Dexia-FondElec Energy Efficiency and Emissions Reduction Fund	Energy	20.0 (sh)		Closed-end venture capital fund that will invest in companies that provide energy efficiency goods and services.
1999	Czech Republic	Brno Waste Water Treatment Plant Upgrading	Water	42.5		Enlarge and upgrade waste water treatment plant and part of the city's sewerage network in ways that meet EU environmental standards. Contribute to reduction of pollution in River Svratka and a man-made lake. Reduce land impact and recover energy by using biogas created during sludge.
1999	Poland	Bydgoszcz Water Supply and Sewerage	Water	26.0		Improve water supply and sewerage services, meeting relevant national and EU environmental, health and safety standards. Reduces pollution to the Vistula and Brda rivers and the Baltic Sea.
1999	Romania	Suez-Lyonnaise MPF Timisoara Water	Water/ Energy	24.7		Multiproject facility to finance water supply, waste water treatment, solid waste disposal, district heating and energy.

Table 5.6
(continued)

Year	Country	Project Title	Sector	Loan (l)/ Shares (sh) millions euro	Co-Financed with other MDBs	Environmental Objectives
1998	Slovenia	Maribor Waste Water Treatment Plant	Water	14.8		Build waste water treatent plan for city and surrounding area to reduce pollution to the Drava river system. Will meet EU and national standards.
1998	Croatia	Zagreb Landfill Rehabilitation	Env.	48.4		Build new waste disposal facilities to improve one of the largest uncontrolled landfills in Europe, in ways to meet EU environmental standards.
1998	Poland	MVV ESCO MPF	Energy	18.7		Multiproject facility to support expansion of ESCOs.
1997	Regional	CGE Municipal Services MPF	Env.	89.5		Multiproject facility to encourage private, municipal financing of environmental services. Has funded waste disposal site in Slovakia.
1997	Romania	Thermal Energy Conservation Project	Energy	40.8		To significantly reduce energy losses and air pollution, improve efficiency of district heating systems in 5 cities.
1997	Romania	Municipal Utilities II	Water	67.9		To upgrade municipal water supply and waste water infrastructure in 10 cities. Will help country meet its commitments to Danube and Black Sea Programmes.

1996	Regional	CGC ESCO Multi-Project Facility	Energy	36.7 (l) and (sh)	Multiproject facility to finance ESCOs. Has financed projects in Slovakia, Hungary, and Poland among others.
1996	Regional	FGG Municipal Services Multi-Project Facility	Energy	41m (l) and (sh)	Encourage private provision and financing of muni and env. services. Has financed Heatco District Heating Project in Slovakia for 3.2m.
1996	Regional	Municipal Services Multi-Project Facility	Water	33.7m (l) and (sh)	RWE Entsorgung Framework Agreement. Finance investments in private provision of muni and environmental services.
1996	Regional	Landis & Gyr Multi-Project Facility	Energy	60 (l) and (sh)	Multiproject facility to support expansion of ESCOs. Has financed projects in Poland.
1996	Regional	Environmental Investment Fund	Municipal	8.6 (sh)	To finance manufacturing, services, development in infrastructure sector.
1996	Regional	Honeywell Multi-Project Facility	Energy	20 (sh)	Establish energy service companies to implement energy efficiency projects.
1996	Slovakia	PBK General Purpose and Energy Efficiency Credit Facility	Energy	14.9	Provide Slovak bank (PBK) customers a facility for low-cost financing for energy efficiency investments.
1996	Romania	Regional Water and Environment Project	Water	20.1	Provide Jiu Valley with constant supply of drinking water.
1996	Romania	Energy Conservation and SME Credit Line	Energy	8.1	Invest in general and energy conservation projects.

Table 5.6
(continued)

Year	Country	Project Title	Sector	Loan (l)/ Shares (sh) millions euro	Co-Financed with other MDBs	Environmental Objectives
1996	Latvia	Riga Water and Environment Project	Water	18.1	EIB	To reduce levels of raw sewage in the Daugava River and improve water supplies.
1996	Hungary	General Purpose Credit Line for EBRD-PHARE Env. and Energy Efficiency Co-funding Scheme	Energy	30		Finance viable private sector projects with environmental and energy benefits.
1996	Croatia	Municipal Environmental Infrastructure Invest. Proj. (Env.)	Water	54.4		Extend and improve sewerage, waste-water treatment and water supply systems for several towns along Adriatic coast. Addresses severe pollution in Kastela Bay and the sea off Pula, and restores quality of bathing water, which will also boost tourism.
1995	Romania	Municipal Utilities Development Proj.	Water	22.2		Rehabilitate and improve municipal water services in five cities.
1995	Lithuania	Kaunas Water and Environment Project	Water	11.9		To improve water and waste-water services in Lithuania's second largest city.

1995	Estonia	Small Municipalities Environment Project	Water	24.0	Short-term water services investments in municipalities other than Tallinn.
1994	Estonia	Tallinn Water and Environment	Water	47.9	Improve water and waste-water services in Tallinn. Loan improves existing water supply and waste-water system, develops new ground water wells, expand waste-water treatment. Improves reliability of water supply services, quality of drinking water, reduces waterborne pollution in Gulf of Finland and Baltic Sea.
1993	Macedonia	Elektrostopanstvo na Makedonija	Energy	35.9	Construct transmission line, support Energy Conservation Programme.
1992	Lithuania	Energy Sector Emergency Investment	Energy	32	For urgent repairs to energy supply facilities to improve supply and end-use efficiency.
1992	Latvia	Energy Sector Emergency Investment	Energy	31.6	Enable urgent repairs to energy supply facilities to improve supply and end-use efficiency.
1992	Estonia	Energy Sector Emergency Investment	Energy	34.9	Enable urgent repairs to power and heating supply and improve energy efficiency.
TOTAL*				979.3 euro/ $1,151.72	

Sources: EBRD information sheets, annual reports, web site (www.ebrd.com/opera/country/). *Exchange rate used is the 31 December rate for each year.

(PBK) in Slovakia, which was enhanced by an additional 3.8m euro interest-free grant from the EU. It has financed subprojects that include a loan to allow a privatized district heating company to replace old boilers with more energy efficient systems, and a loan to help an industrial company replace its old air compressors.

Among the other energy projects in table 5.6, energy conservation is a component of the broader project. For example, the Bank's 1993 project in Macedonia had several main goals aimed at helping the government restructure and strengthen its power sector. One major goal was to construct a new transmission line to help the capital city, Skopje, achieve a more reliable supply of energy, which reduces the demand for electricity imports and saves scarce foreign exchange. Another goal emphasized energy conservation, by supporting the Government's Energy Conservation Programme. This involved the EBRD helping enterprises preparing for privatization to reduce their energy costs, in part to enhance their attractiveness to future investors.[94] The project also financed an energy conservation unit in the Ministry of Economy, to help implement energy efficiency plans.

Other non-ESCO projects include the Romanian Energy Conservation project, which provided financing for the rehabilitation of district heating systems in five different cities, and the three Baltic emergency energy sector loans, which were driven by the need for these countries to cope with sharply reduced fuel imports from Russia. All three loans contained energy efficiency components, as well as other energy supply components. The Estonian loan, for example, financed stations to meter gas imports from Russia, to help set up trade with Russia on a normal, commercialized basis; the completion of facilities to import and store heavy fuel oil to help Estonia meet its needs for several winters; spare parts to restore electricity and power output for power generation and heat production; and measures to improve energy efficiency in industry, residences, and energy utilities.[95]

The EBRD's water projects, in turn, are produced by the Municipal and Environmental Infrastructure Unit. These projects resemble the World Bank's, in the sense that their goals are to improve wastewater collection and treatment, as well as drinking water quality. Like the World Bank, the EBRD has financed projects in the Baltic states that are in areas deemed hot spots by the 1992 JCP. The Romanian projects, although not sought out by the Bank as a means to clean up tributaries to the Danube River, are nonetheless in areas where the quality of water and wastewater services was very poor.

The EBRD's water projects differ from the World Bank's in two ways. First, they do not include the additional, small "environmental management component" found in the World Bank's projects. The approach of the EBRD projects tends to be narrower, with a greater emphasis on the corporatization of the municipal entity and on encouraging the private sector whenever possible. For example, the Maribor project was constructed as a "build-operate-transfer" (BOT) operation between the city and a private operator. A second distinction between the EBRD and World Bank projects is loan size; the average EBRD loan for water projects is around 24 million euro ($30 million), versus $5 million for the World Bank. This difference in loan size is a result of the fact that the EBRD is less willing to make small, individual loans, which has led it in some cases to make one loan that is in turn provided to a number of separate municipalities.

Generally, the projects in table 5.6 are distinctly more banklike than those in the World Bank's portfolio, in the sense that they do not include the scope of environmental management and capacity-building components of the World Bank's projects, nor do they cover areas such as forestry, or the broader category of sector loans that emphasize more macroeconomic policy changes such as the removal of energy subsidies. Yet, since most of the EBRD's projects are the work of specific "green" units within the Bank, they do reflect the Bank's desire to pursue activities that other bankers—within the EBRD, or in private sector banks—may not pursue. For example, in developing a niche in ESCO financing, the EBRD is designing projects that are typically much smaller than the average MDB loan, and the ESCO sector is not one that commercial banks have much experience in. The water projects, in turn, do not emphasize a broader watershed management approach, as do the World Bank's projects. Yet some of them—namely, the Baltic Sea projects—do address regional areas where water pollution reduction is a priority.

As is the case of the World Bank, the projects in this portfolio emphasize environmental issues relatively more than other projects in the same sector. For example, the EBRD's broader set of energy projects includes loans to construct or reconstruct electricity networks, develop gas storage and distribution, and finance the modernization of gas station networks. Also similar to the other two banks, the EBRD has some energy projects in its broader portfolio that have environmental benefits, but these are not the primary or significant components or goals of the projects. The modernization and rehabilitation of power facilities, for example, tend to result in energy savings and can also avoid the need for additional power plants.

Role of Recipient Demand

EBRD projects are often initiated when Western companies or CEE countries or cities contact the Bank with a suggestion for a project. However, the EBRD still differs significantly from a private sector bank in the degree to which it builds policy conditionality into all of its loans and requires private sector borrowers to address certain environmental issues or standards. For example, a loan to EGIS Pharmaceuticals in Hungary contains environmental covenants to monitor the implementation of the company's environmental compliance plan to reduce its wastewater discharge, its volatile organic compounds emissions, and any soil and groundwater contamination. This is the area where the work of the Bank's environmental appraisal unit often shows up, since it can add environmental conditionality into loan covenants. The Bank's environmental portfolio is probably relatively less demand-driven than other areas of its lending. For example, many municipal institutions or private enterprises in Central and Eastern Europe are unfamiliar with the concept of ESCOs, which must be sold to them by the EBRD, or the energy companies it finances. Even the first Western companies setting up the ESCOs in CEE had to be persuaded to move into the Eastern European market and to contribute some of their own finances to the projects.[96]

In addition, the municipal infrastructure projects use sovereign guarantees to a greater degree than other departments at the Bank. Sovereign-guaranteed loans give Bank staff more leverage to build conditionality requiring policy changes than they have with private sector loans. Such policy changes, however, are bottom-up—often at the municipal, utility, or company level—versus the central government. Examples of this would be conditions requiring municipalities to raise tariff levels as part of a water and wastewater treatment project, or requiring towns to penalize companies that are not meeting pollution emission standards.

Finally, loans developed in the municipal sector initially had much longer lead times and higher preparation costs than loans developed in other areas of the Bank, particularly in first years of lending for municipalities. This reflected the challenges the Bank faced in devising loans for municipal agencies first gaining their autonomy from central governments and that had little experience in borrowing or in operating their agencies on a market basis. More banklike financial institutions would stay clear of such challenges, particularly given the fact that most municipalities were not creditworthy enough to borrow on their own. Given that this sector of lending also had more of a development angle than the EBRD's work in other areas, it attracted a great deal of bilateral grant financing and other forms of donor assistance, often through the Project Preparation Committee.[97]

On Implementation
Neither the EBRD nor the EIB provides the depth of information available on project implementation from the World Bank. As a result, assessing the implementation of EBRD and EIB projects is difficult. This section draws mainly on interviews with central and municipal government officials, MDB staff, and environmentalists, and on visits to various CEE projects to meet with the project leaders. This research has shown that while—as in the case of the World Bank—there are no projects that are considered to be major environmental disasters, there are a number of the projects where implementation does not occur as planned, and there are specific projects that concern NGO watchdog organizations.

The main environmental group in the region that acts as a watchdog for MDB activities is CEE Bankwatch Network, whose members represent domestic environmental groups, as well as regional representatives of international NGOs such as Friends of the Earth.[98] CEE Bankwatch has been supportive of some of the EBRD's activities, particularly work done by the Bank's Energy Efficiency Unit. As CEE Bankwatch Network's Energy Project Coordinator Petr Hlobil noted, "Only the EBRD has adopted a progressive approach toward energy efficiency; it actively supports energy service companies and credit facilities for energy-efficiency projects."[99]

At the same time, looking at the broader EBRD portfolio, as noted above, environmentalists have been critical of a number of project choices considered or funded by the EBRD, such as the Mochovce nuclear plant and the ZSNP smelter (mentioned in chapters 2 and 4, respectively). In terms of implementation issues, environmentalists have been most disparaging of the inability of the EBRD's nuclear safety account (NSA) to result in the closure of any high-risk nuclear reactors in the region. After organizing to fight against passage of the 1995 proposed Mochovce loan, environmentalists have also unsuccessfully fought against a later proposal to finance the Rovno and Khmelnitsky partially built 1000 MW VVER nuclear plants in the Ukraine. The Ukrainian government intended to complete these Soviet-designed reactors to replace two units at the Chernobyl nuclear plant. However, in February 1997, an independent panel commissioned by EBRD concluded that project was not a least-cost option, given declining demand for energy in the Ukraine and significant scope for energy conservation and demand-side management. Nonetheless, the Bank moved forward in the loan process, while environmentalists argued that the Bank should stop financing this project, given serious concerns about the nuclear plants' safety.

In Bulgaria and Lithuania, in turn, NSA-supported financing, which was aimed at providing short-term improvements to nuclear plants in return for the plants' closure, resulted instead in government maneuvers to extend the life of particular plants beyond dates desired by the EBRD and donor countries. In Bulgaria, while a 1993 agreement between the country and EBRD called for the government to shut down two of its oldest Kozloduy nuclear reactors by the end of 1998 and to shut down an additional two by 2000, the government reneged and decided to extend the closure dates to 2006 and 2008, respectively. The EU was later able to exert pressure for the country to move the closure dates forward, as noncompliance was seen as damaging to Bulgaria's prospects of joining the EU.[100]

The situation was somewhat similar in Lithuania, where an agreement was signed with the NSA in 1994. Under the agreement, the government would not extend the life of Ignalina nuclear plant's two units beyond the time when their fuel panels would need to be replaced.[101] However, under a safety assessment that was part of the agreement, a group of independent safety experts commissioned by the Bank and the Lithuanian government recommended that the plant's two units should not be restarted after a 1997 shutdown for maintenance, until numerous safety issues were resolved in its operation and design. While the EBRD pressured Lithuania to close Ignalina, the government balked. Pressure from the EU finally had an impact on the government, which agreed in 1999 to shut down one reactor by 2005 and to decide what to do with the other, older reactor by 2004.

Turning to the projects in the Bank's environmental portfolio, to date there has been little assessment of credit-line-related activities, but other investment projects appeared to be performing reasonably well, despite the existence of various implementation difficulties. In the water sector, for example, the EBRD's projects face some of the same challenges as the World Bank's, namely, municipal water companies struggling to make profits (and repay loans) in the face of sharp declines in water consumption. As in the World Bank's projects, the EBRD water companies have had a difficult time raising tariffs, particularly in cities or towns facing increasing levels of unemployment and poverty. These projects also faced their share of various—often small—implementation problems.

The Tallinn Water and Environment project, for example, was generally seen as successful in reducing the discharge of untreated wastewater into the Baltic and cleaning up drinking water. The project financed rehabilitation of Tallinn's water treatment plant, existing ground water wells, the booster stations and network, and old sewerage pumping stations. It also financed the replacement of old consumer water meters as well as the installation of new meters and upgraded the city's activated sludge treatment plant to meet HELCOM recommendations.[102] However, it did face unexpected delays when slow parliamentary approval of the project put the project nine months behind schedule. Also, the costs of investments and upgrading planned by the project were higher than anticipated, but the Tallinn Water Company was able to raise the additional funds from local banks. According to Tallinn water officials, treated wastewater going into the Baltic Sea was by the late 1990s two to three times better than in 1993. Drinking water was also significantly cleaner after treatment, but further investments were needed in the water network to get the higher quality water to the consumer.[103]

In Kaunas, Lithuania, the EBRD's project was applauded by environmentalists, who had lobbied hard for the construction of a wastewater treatment plant in the late 1980s. With a population of around 403,000, Kaunas is the second largest city in Lithuania, and before the

EBRD project it had no wastewater treatment and was responsible for around 90 percent of the untreated wastewater discharged in the country. As such, Kaunas was one of the hot spots identified under the 1992 Baltic Sea Environmental Action Programme.[104] The project financed the first phase of the construction of a wastewater plant that included mechanical treatment with chemical flocculation, while biological treatment of waste-water was planned for the medium term.[105] It also financed new pumping stations and a new sewer network, among other components. Finally, the project, like most other MDB water projects, "twinned" Kaunas Water with a Western water company (in this case Stockholm Water Company), to help improve its management and implementation.

The Kaunas project had a slow start and was delayed for over a year, mainly owing to difficulty meeting covenants on tariff increases. The initial project design, inherited from the Soviets, was also determined to be too big, after assumptions about water consumption were seen as too high. One actor involved in the project also noted that the central government was slow in transferring its share of project financing, while others complained that a number of the bilateral actors involved in components of the project (such as PHARE) provided poor consultants. Some of the project delays were also attributed to the constant political changes in the city, given that there were eight different mayors from the time the project was agreed to mid-1998.[106] As in other municipalities, tariff increases proved too painful to be implemented as planned. The municipality also decided not to implement the planned construction of a potable water treatment plant because it was too expensive.[107]

EIB

General Approach

The EIB's entry into CEE is a product of the desire by the EU and its member states to respond to the collapse of communist regimes in the east while preparing to offer membership to a number of its former Soviet-bloc neighbors. The EIB was "invited" by the European Council to initiate lending to Poland and Hungary in October 1989, and the money began to flow the following year. The Bank received a mandate to lend 1.7 billion ECU to the region in 1990–1993; 3.0 billion ECU in

1994–1996; and 3.0 billion euro for 1997–2000. In the latter period, as noted above, the Bank also set up an additional 3.5 billion euro preaccession lending facility for the ten CEE countries and Cyprus, which differed from the lending mandates in that it did not require EU member state guarantees. The amounts continued to grow, with the Bank authorized to lend an additional 8.7 billion euro to the ten applicant countries, Albania, Bosnia-Herzegovina, and FYR Macedonia for 2000–2006, as well as an additional 8.5 billion euro for a renewed preaccession facility for the period 2000–2003.[108]

The scope of the Bank's coverage in the region has also expanded over the past decade. During the first period, lending was authorized to Bulgaria, Czechoslovakia (later separated into the Czech Republic and Slovakia), Hungary, Poland, and Romania. In the first three years of activities in CEE, just over half of the Bank's loans went to Poland and Hungary. Lending soon expanded to include Albania, Estonia, Latvia, and Lithuania, while lending to Slovenia began in 1993 under a separate financial protocol with the EU and lending to FYR Macedonia resumed in 1998.[109]

The EIB's Portfolio of Activities

In general, the Bank's lending in CEE has a very heavy emphasis on infrastructure work, particularly in the area of trans-European networks, or projects that better link the region to Western Europe. Table 5.7 reveals that transport is the Bank's single largest sector of lending to the region, with 49 percent of total commitments, and most of these loans are used for building roads and highways. The other well-represented sectors include telecommunications, loans in the banking/financial sector, and energy lending, mainly for power generation oil and gas. The EIB tends to undertake similar types of projects in different countries. For example, in most countries in the region, it is involved in air traffic modernization, highway construction, and telecommunications modernization. All countries in the region have also received global loans for financial intermediaries to on-lend to small- and medium-sized enterprises, in areas that may include energy efficiency or environmental protection, as well as tourism. Clearly, the EIB's range of lending is much narrower than those of the other two MDBs.

Table 5.7
European Investment Bank CEE commitments by sector, 1990–1999, in millions euro

Sector		Sector Total	% of portfolio
Transport:		5321	48.7
Roads/Highways	2984		
Railroads	1698		
Airports/Aircraft	325		
Other (includes port rehabilitation, metro)	314		
Telecommunications		1517	13.9
Banks (global loans, credit lines, financial sector reform)		1457	13.3
Energy:		1334.5	12.2
Power Generation	609		
Oil and Gas	680		
Heat Restructuring/Conservation (District Heating)	45.5		
Manufacturing		517	4.7
Misc. Env.		183	1.7
Water		144	1.3
Misc.		445	4.1
TOTAL		10.9 billion euro	100

Sources: EIB, "Summary of Signed Operations in Central and Eastern Europe," June 1996; EIB press releases, various dates, EIB Annual Reports, 1990–1999.

Even in common sectors, such as transport, there are differences in the three banks' emphases. The EIB and EBRD, for example, have financed more projects to construct new highways and road in the region, while the World Bank emphasizes road maintenance versus new construction. Environmentalists, as discussed below, have been highly critical of the EIB's work in road construction, arguing that a sustainable development approach would warrant a greater emphasis on public transport, such as railroads. The EIB's work in road construction, in turn, echoes the EU desire to finance "trans-European networks."

Approach to Addressing Environmental Issues

Chapter 4 has shown how EIB staff have limited incentives to pursue environmentally oriented projects beyond traditional infrastructure projects that have environmental components. The Bank's project portfolio supports that analysis. Put into the language of the continuum from chapter 1, the EIB's activities fall largely in the "lighter green/minimal activity" area. A handful of energy and water projects can be argued as belonging to the category of projects that go beyond minimal environmental actions and have significant but not primary environmental objectives. A single project has a primary environmental goal; that is, a 13 million ECU loan to help finance reforestation of abandoned state-owned agricultural land in Poland. The EIB cofinanced this project with the World Bank, which took the lead in project development.

Unlike the other two MDBs, the EIB has spent little time and effort on broader environmental policy issues. As the Bank itself noted in the early 1980s, "The EIB has no role, of course, in the broad political choice of investment priorities (e.g. the arguments for investments in rail as against road, nuclear as against other power stations, etc.)."[110] The EIB has been a peripheral player in the policy networks that have developed around the Environment for Europe process. However, it has been relatively more active in regional efforts to address environmental degradation of the Mediterranean and Baltic seas. It worked with the World Bank to set up in 1988 the Mediterranean Environmental Technical Assistance Programme, which carried out priority investments in the area. In 1992 it also began to participate in the Baltic Sea Joint Comprehensive Action Programme, under which it has financed two wastewater treatment projects in Central and Eastern Europe, as well as several others in northern Germany and Sweden.[111]

Table 5.8 lists EIB loans for projects that have primary environmental goals or significant environmental components, which total 501 million euro for ten projects, or under 5 percent of its total portfolio. Although the Bank describes a number of its global loans as being used to finance projects that have goals of energy savings or environmental protection, it does not release information on how the global loans are on-lent, so there is no way to know what percentage of those loans are actually used for green projects. Even the municipal loans in table 5.8 with an environ-

Table 5.8
EIB projects with primary environmental goals or significant environmental objectives in CEE: 1990–1999

Year	Country	Project Title	Sector	Loan/ Shares millions euro	Co-Financed with other MDBs	Environmental Objectives
1999	Latvia	Upgrading env./municipal infrastructure	Munic.	20		Support mainly environmental investments by medium and small sized municipalities. Focus is on wastewater, water, solid waste, urban infrastructure. Will have positive impact on Baltic Sea area.
1999	Lithuania	Upgrading env./municipal infrastructure	Munic.	15		Funds for a number of towns to upgrade environmental and other public infrastructure.
1999	Lithuania	Upgrade waste water plant	Water	6		Assist city of Panevezys in upgrading main waste water treatment plant and sewer network.
1998	Bulgaria	Protect Black Sea coast and Danube River Bank	Water	25		Finance around 30 schemes to repair damages caused by landslides on Black Sea coast, and to halt erosion along the Danube.
1998	Hungary	Improve urban transport, sewerage, solid waste disposal, amenities	Munic.	110		Project will add flue-gas treatment to solid waste incinerator, improve city parks and baths, help eliminate traffic congestion, replace obsolete tramcars.
1996	Latvia	Riga Water and Environment Project	Water	15	EBRD	Upgrade, rehabilitate water supply, sewerage systems in and around Riga to help with reduction of pollution to Baltic Sea, in line with Baltic Sea Joint Comprehensive Environmental Action Programme.

1995	Czech Rep.	C'EZ I	Energy	200		Install desulphurization equipment in six thermal power plants to bring them into compliance with international standards. FGDs will reduce SO_2 emissions by more than 90% in this Black Triangle region.
1994	Estonia	District Heat Rehabilitation	Energy	7	World Bank	Rehabilitate district heating system in Pärnu and Tallinn as part of the World Bank's District Heating Rehabilitation Programme. Finances equipment to convert oil-fired boilers to wood and peat, pumping stations, water treatment facilities, pipe insulation.
1994	Poland	Warsaw Sewerage Treatment Plant	Water	45		Construct new waste water treatment plant south of Warsaw on left bank of the river Vistula. Helps reduce pollution into the Baltic Sea.
1993	Poland	Forestry Development	Forests	13	World Bank	Afforestation program in parts of Poland. Helps finance the planting of around 17,000 hectares of trees between 1993–95.
1992	Bulgaria	Energy I	Energy	45	EBRD	Help finance completion of lignite fired generator fitted with flue gas desulphurization equipment at Maritsa East II.
TOTAL*				501 euro/ $605		

Sources: EIB, "Summary of Signed Operations in Central and Eastern Europe," June 1996; EIB press releases, various dates; EIB Annual Reports, 1990–1999.
*Exchange rate used is ECU/Eur dollar rate from 31 Dec. each year.

mental focus are not broken down so that an analyst (or the public) can know how the money is actually being used. Around 40 percent of this amount was for a single project—the CEZ I project in the Czech Republic, which financed the purchase of expensive desulphurization equipment for six Czech power plants. The Bank's energy project in Bulgaria also financed the purchase of FGD equipment for a power plant (in this case the Maritsa East II plant) and was part of a larger EBRD project there.

These projects have environmental benefits because they result in a reduction of SO_2 emissions. The EIB's CEZ project was similar to that of the World Bank in its emphasis on installing FGD equipment in power plants in northern Bohemia. The EIB project financed the installation of this equipment at six power plants and estimated that as a result SO_2 from these plants would drop by 93 percent and particles would decline by 50 percent. The Bulgaria project, in turn, involved cofinancing with the EBRD for the completion of the Maritsa East II plant, including retrofitting with FGD, expanding the plant's ash disposal pond, rehabilitating its water supply system, and purchasing equipment to measure air quality. The EIB provided 40 percent of the project's finance, and the EBRD provided 35 percent. The project documents of the EIB, unlike similar project documents from the other two banks, do not stress environmental issues in the rationale for funding. In the case of the Bulgarian project, for example, the EIB emphasized the importance of increasing the safety and efficiency of Bulgaria's electricity capacity in order to provide more electricity.

Several of the other projects in this table are cofinanced with one of the other two MDBs. These are cases where the projects were developed largely by either the World Bank or EBRD, and the EIB came on board to finance a particular component. The Estonia district heating project, for example, was undertaken as part of the World Bank's District Heating Rehabilitation Program in Estonia. The EIB project focused on rehabilitating the district heating system in the city of Pärnu, although it also included a small component for rehabilitating the district heating system in Tallinn. The goal of the EIB's project was to reduce fuel costs to the company through fuel substitution and other energy savings measures, as a way to help Pärnu's district heating enterprise improve its performance in the face of declining demand and higher energy costs. The result would be significantly lower emissions of SO_2, NOx and CO_2, although it would also produce a minor increase in particulates emissions. The EIB's loan financed a series of studies, equipment, implementation, and operation of facilities that converted two stoker grid boilers to be able to burn peat and woodchips with better combustion conditions. The "new fuels" are cheaper than the heavy fuel oil normally used in these facilities.

The EIB's component of the World Bank–designed Polish forestry project involved a loan to Poland's environment ministry to help finance the planting of trees on abandoned state-owned agricultural land throughout Poland. It was part of a five-year program, supported by the World Bank, to rebuild and extend damaged forests, in part to ensure supplies for the country's wood-processing industry. The Riga water project, in turn, cofinanced the upgrading and rehabilitation of water and wastewater systems in the Riga area as part of a larger EBRD project. This project was one of the hot spots identified by the Baltic Sea Joint Comprehensive Environmental Action Program. Finally, the EIB's wastewater project in Warsaw was also being undertaken under the auspices of the Baltic Sea program, as another hot spot requiring urgent action.

Role of Recipient Demand, and Implementation
The EIB is widely acknowledged by government officials, environmentalists, and project managers in CEE to be the most demand-driven of the three MDBs. In addition, its loans contain the least conditionality and offer the cheapest rates, without the up-front fees charged by the other MDBs. The expectations developed in chapter 4 are borne out by analysis of the Bank's work in CEE; that is, the absence of any green bankers, the small size of the EIB staff, and the dearth of Bank-sponsored policy work all result in the Bank having a less active approach to project identification than the other banks. There are even a number of examples of projects turned down by the other two MDBs for economic or environmental reasons that the EIB has decided to finance, leading NGOs in CEE to label the EIB a "bottom-feeder."[112]

Of the three MDBs, the EIB is also the most impenetrable on the issue of project implementation, so it is difficult for the public to know how well the Bank is supervising projects and how projects are performing. The Bank's small evaluation unit has noted that project monitoring is often weak or nonexistent.[113]

NGOs following the Bank's activities in CEE have a number of complaints about the Bank's lack of transparency and its prickly, often defensive relations with them.[114] They have grown increasingly critical of the EIB's work in the transport sector, given its emphasis on building new roads over rehabilitating and maintaining old roads.[115] NGOs argue that the Bank would be more sensitive about promoting sustainable development if it focused on public transport instead of roads, which will increase road traffic more than rail transport. They point out that the enormous shift toward road transport and car ownership occurring in the region in recent years will result in increasing levels of air pollution and loss of biodiversity, and they call for greater emphasis on the management of the demand for transport and urban planning.[116]

NGOs launched a campaign to fight the EIB's M3 toll road in Hungary, which extends from Budapest to the Ukrainian border, passing through a relatively underdeveloped region of Hungary.[117] They questioned the need to build a new toll road in an area where traffic levels are low and where there was no connecting road on the Ukrainian side of the border. The extension of the M3 seemed to be a road going nowhere. Hungarian environmentalists argued that financing would be better spent in road maintenance. Indeed, the World Bank was approached first to finance this project, but it refused, arguing that Hungary's road maintenance expenditures were only 25 percent of the total needed to prevent further road deterioration, and that Hungary should spend more on maintenance and less on new roads.[118] Even the EBRD had noted that the far eastern sections of the road were not commercially viable as a toll road, given its location in a poor area of Hungary. The EIB, however, argued that the short section of the M3 that is covered by its loan was in fact economically justifiable. Its economic appraisal, however, was not available for public scrutiny.

The project itself has been supported as a major link in the planned "trans-European network" extension, as part of a Trieste-Ljubljana-Budapest-Lvov-Kiev corridor that was identified at the March 1994 Second Pan-European Transport Conference held in Crete. Pressure for the M3 road has also stemmed from members of parliament from Eastern Hungary, Hungary's trade ministry, and a "concrete-lobby" of construction firms, trade unions, local mayors, and Socialist party groups.[119]

As for the other projects on the list, as of the late 1990s, the Warsaw water project (signed in 1994) was stalled, and implementation had not begun. The project faced difficulties regarding the ownership of land where the new wastewater treatment plant would be built, since property rights of the proposed site were contested.[120] If it eventually moves forward, the plant could play an important role in cleaning up the Vistula River, into which most of Warsaw's sewerage flows, untreated.

Conclusion

The MDBs are major donors in CEE, and their approach to addressing environmental issues in their work can have an important impact on policy reform and infrastructural development in the region. This chapter has described the MDBs' activities in CEE and the ways in which those activities reflect how demand-driven and banklike each institution is. The World Bank, the least banklike and with the largest network of internal green bankers of the three, has had the flexibility to undertake a broader array of environmental projects and policy activities than the other two banks. Its role in the GEF gives it an additional portfolio of projects not found at the other two banks. The EBRD, in turn, with its three environmental units, has created a more limited niche for itself in environmental financing in the areas of energy efficiency and municipal infrastructure. Although its activities are narrower in scope than the World Bank's, its green lending volume, at least in terms of commitments, is very close to World Bank levels. Finally, the EIB's portfolio shows the least evidence of a conscious pursuit of projects with significant or primary environmental components or goals. Its environmental projects reflect more its position as a cofinancier of projects designed largely by the other two banks, rather than any explicit initiative to search for such projects.

The chapter has shown that the more demand-driven the bank, the less likely it is to be involved in efforts to help governments reform their policies in ways that benefit the environment. The catch, however, is that the more creditworthy a country is, the more options it has in finding private sector financing for infrastructure and other activities, and the less leverage the MDB has. There is an inverse relationship between an MDB's leverage and a country's economic health.

This chapter has also highlighted the fact that the greenest projects tend to be designed by bankers or units with the specific task of pursuing such projects. Even in the same lending sector there are often visible differences between projects that involve these people and projects that do not. Environmental impact assessment units, in turn, tend to work as the banks' internal environmental watchdogs. Although banks like the World Bank and EBRD have become more environmentally conscious over time, the pace and process of "mainstreaming" these ideas remains slow and uneven. NGO watchdog groups point out inconsistencies and gaps in bank policies and strategies, as well as shortfalls where the banks have paid lip service to broader issues, such as the impact of power projects on greenhouse gas emissions and global climate accords. Projects have also faced a variety of problems during the implementation process, and all three banks have struggled to meet their goals in the region. Ultimately, it is often difficult to measure the actual environmental impact of MDB projects in countries, while it is easier to point out areas where the projects contradict policy goals, or where lofty ideas do not take root. At the same time, the chapter has highlighted some successful or innovative projects and ideas that make use of the institutions' comparative advantage, such as the EBRD's work with ESCOs. Adding up the banks' actions and activities, the environmental impact in CEE seems far more modest than the rhetoric would lead one to believe, but less disastrous than some of the broader NGO protests against IFIs might lead us to expect.

6

Conclusions

International financial institutions such as multilateral development banks face mounting criticism of their performance at the same time their mandates are expanding and demand for their services is increasing. The World Bank, along with the World Trade Organization and the IMF, have been singled out by antiglobalization activists in increasingly violent protests against the institutions' perceived negative impact on the poor, on the environment, and a host of other issues. It is only a matter of time until such attention turns to the EIB, which is far less transparent and accountable than the World Bank. MDBs are being scrutinized by some as loyal servants of states who continue to remain the dominant actors on the international scene, and by others as independent actors that can shape state interests and sometimes even tell states what to do. This book has argued that institutional behavior is not an "either-or" issue with respect to how an international institution's actions reflect member state interests versus the institution's own interests and capacity, but rather, a more complex "when and how" issue. The ability of MDBs to take on new mandates, such as the environment, depends on the strength of shareholder commitment, whether or not new goals conflict with pre-existing aspects of institutional design and incentive systems, and the leverage the banks have in "selling" activities embodying the new mandates to recipients. These issues therefore have important policy implications in debates about how international institutions should be designed and what types of factors can improve their performance in a more interdependent world.

There is no shortage of policy prescriptions about how to improve these institutions' performance; what is striking is the range of proposed solutions—from abolishing the institutions to narrowing their goals, or expanding their goals to the point that they are no longer financial institutions at all but another form of grant-making donor. With the exception of calls for closing MDB doors, the policy prescriptions propose cures that would make the institution either more or less banklike. Yet, as this book has shown, movement too far in one direction or the other has its costs, with the more banklike MDBs having fewer institutional incentives to pursue environmental projects and the less banklike MDB struggling more to juggle multiple goals and facing renewed scrutiny where lofty rhetoric has little or no tangible results. Not enough attention has been paid by MDBs critics and supporters alike to thinking about how these institutions can better balance their dual personalities.

This study shows that the ways in which MDBs react to new environmental mandates are significantly shaped by the interaction between shareholder preferences, organizational dynamics and recipient demand. It argues that attempts to understand the environmental behavior of these institutions requires an analysis of what happens at three stages of the policy process, since any one of these can be a source of innovation or inertia. Tracing the process of how the institution receives new policy objectives, how it institutionalizes them, and how it translates these ideas into specific activities shows more precisely the ways in which external politics and internal organizational characteristics and incentives interact to produce outcomes.

New mandates tend to enter the MDBs through external channels, reflecting the interests of major board members. In the case of the World Bank, environmental NGOs played a significant role in how the United States defined its interests with respect to the environment. For the EBRD, the interests of major donors were enhanced rather than directly driven by NGO support. The book also shows how all three banks share the characteristic of diffuse governance structures. This makes it difficult for new mandates to "infect" the institution, and therefore new policy demands are constrained by organizational mission, design, financial tools, and a variety of other factors. Added together, these factors tend to reflect how demand-driven or banklike the MDB is, which highly influences its ability to translate new policy goals that differ from more traditional MDB attention to economic development issues. Finally, even if the institution is able to offer a certain set of green activities, its ability to "sell" them to recipients depends on recipient demand for these activities and the leverage the MDB has to influence the borrower's choice of projects and programs.

At an MDB like the EIB, the story to date essentially stops at the first stage of the policy process. Weak donor commitment to the Bank's environmental activities results in few if any organizational innovations to improve the incentives of its staff to actively seek out projects for environmental reasons. Although the Bank says that it applies European Union environmental standards in its lending, this study has shown that oversight and transparency are weak within the Bank, and that these standards themselves can be of mixed strength.

The World Bank and EBRD, in turn, face the challenge of how to institutionalize and implement the strong environmental mandates given to them as a result of pressure from major shareholders backed by environmental NGOs. For the EBRD, this means squaring its environmental mandate with its private sector emphasis, since the more demand-driven an MDB is, the less ability it has to seek out projects with environmental goals that are not of interest to recipients. It has tackled this challenge through its environmental procedures and through a few discrete units set up within the bank to design projects that emphasize environmental goals in ways that dovetail with its interests in promoting private sector development. The World Bank has had more flexibility than the EBRD in many ways, given that its mission and institutional design have made it the least demand-driven of the MDBs and the most oriented toward influencing recipient governments to adopt projects that come with conditionality effecting broader policy changes.

Even where an MDB is able to undertake environmentally oriented projects, the cases of the World Bank, EBRD, and EIB in Central and Eastern Europe highlight some limits to being a darker green MDB. An MDB has most leverage in selling certain types of policy ideas through its work either where the goals of the bank and client overlap, or where the recipient has no alternative sources of financing that require less policy conditionality. The World Bank has been most influential in CEE countries in the earlier stages of the transition process. As CEE countries grow economically stronger and have access to international capital markets, they have less incentive to agree to World Bank loans. In Poland, Hungary, and other preaccession countries, World Bank lending volume is shrinking, and the Bank's role will increasingly turn toward advice and technical assistance versus projects and structural adjustment lending. To the extent that these countries can raise funds on international capital markets, or have access to various EU sources of grant funding or the EIB's large pots of cheap funds with less conditionality attached, the World Bank will have a tougher time finding new business in the region.

The preceding chapters have also shown the complexity of conceptualizing the "environmental behavior" of an institution, since it can encompass behavior ranging from a minimal response emphasizing mitigation measures for projects seeking to promote other—possibly conflicting—goals, to a more active response that aims specifically to achieve an environmental end. This is an important point because it focuses attention on the types of projects MDBs are set up to undertake, given how bank-like their design and mission are. In other words, if an MDB is designed to be demand-driven, it is unrealistic to expect its staff to spend time identifying and designing environmentally oriented projects that are not sought out by its clients. We should instead expect bank officials, at minimum, to follow a clear set of environmental due diligence procedures to avoid financing environmentally destructive projects, and to improve project design to reduce potential negative environmental effects. In addition, there is a set of more traditional MDB activities that happily meet the dual objectives of environmental improvement and economic development—such as projects that seek to modernize or build more efficient energy facilities, or sectoral restructuring and structural adjustment work that calls for a reduction in energy subsidies. Whether or not these activities are old wine poured into new bottles—examples of institutional adaptation versus real learning—tends to reflect the involvement of the banks' green bankers.[1] In fact, most of the relatively darker green projects in CEE were found in the energy and water sectors, traditional areas of MDB activity. Yet projects with primary environmental goals or significant environmental components could be distinguished from more traditional project designs.

The environmental benefits of MDB activities are also often difficult to quantify. This is particularly so for MDB contributions to policy development and sectoral reform through nonlending activities such as technical assistance and agenda-setting exercises. Exercises in capacity building, policy prioritizing, and training can be highly important but impossible to measure.

In evaluating outcomes, in terms of whether projects are implemented as planned, the projects analyzed here illustrate why project documents are essentially guides that are rarely followed as written, and also show some of the types of unanticipated obstacles or delays projects can face in the implementation process. This underscores the point that it is sometimes difficult for an interested public to determine the actual environmental impact of a project.

An examination of the environmental management components of the World Bank's Baltic water projects, for example, showed how some components fizzled out, while others—such as the pig farm study in Siauliai, Lithuania, discussed in chapter 5—may have a significant effect on reducing the amount of pollutants entering local water sources. An analysis of available data, backed by interviews with more than 100 MDB officials, environmentalists, donor and recipient government officials, and other actors, showed that there were no environmentally disastrous MDB projects in CEE, compared with the types of projects in other parts of the world that fueled the NGO campaign against the World Bank in the 1980s, a campaign that has picked up speed in recent years amid a new round of protests against the costs of globalization. But disastrous projects aside, MDB environmental performance in CEE is still mixed at best, given the number of projects where the banks have not followed their environmental procedures, project goals are dubious in light of the Bank's stated policies, or the portfolio overall does not address the larger issues, such as its impact on producing greenhouse gases.[2]

The EIB has made by far the smallest environmental contribution of the three MDBs to date, but it is likely to face increasing pressure over time to improve its environmental performance. In particular, it will have to help accession countries comply with the EU's environmental *acquis communautaire,* in support of the accession process. Stirrings of anti-EIB publicity materialized in 1999–2000, and the first critical report by a government agency from an EIB donor country has been published. As Britain's Department for International Development noted regarding the EIB's external lending operations: "There is no agreed statement setting out how the Bank's proposed activities will contribute to the EU's international development aims and external regulations. There is a need for a clear set of objectives together with associated performance indicators. Without these it is difficult to judge the effectiveness of the Bank's operations and whether it makes good use of the grant resources it receives from the EU."[3] Perhaps the writing is on the wall, as member states realize that what they are asking the Commission and the EIB to do may not always match.

Theoretical Implications

The arguments made here challenge various schools of thought that seek to explain the behavior of international institutions. This book joins the ranks of historical institutionalist approaches in pointing out how neorealist and neoliberal institutionalist theories of institutional behavior can be static and poorly equipped to account for what happens when institutions are given somewhat conflicting demands by their member state owners. It also provides evidence for institutionalist arguments about the sticky behavior of institutions; that is, even if member state preferences change over time, institutional structures, procedures, and incentives may adapt more slowly or not at all. In addition, these cases argue for the importance of looking more closely inside the institutions to determine the ways in which their organizational design and incentive systems shape behavior. Put differently, this suggests the importance of bringing back some of the lessons of organizational theory into the way international relations scholars study international institutions.

Generally, existing theories on international institutions tend to focus on one of the three stages of the policy process. What is needed is a better integration of international relations theories emphasizing the importance of shareholder politics and preferences in shaping institutional behavior with organizational and institutionalist theories that focus attention on how what happens inside the black box of the institution can shape behavior. There is also great scope for further research on the mechanisms by which new ideas "infect" institutions, as well as how organizational factors shape knowledge entering international institutions.

Policy Implications

Three sets of policy implications arise from these arguments. First, NGOs and other actors that seek to change the behavior of MDBs may be most effective when they attempt to influence the MDBs' boards of directors.

There is little evidence that the institutions themselves are sources of significant change in environmental behavior, although institutional actors have been important in shaping how the broader mandates are translated into policies and procedures, for example, through the development of environmental assessment procedures. In the three MDBs examined here, the major policy changes have external sources. Bank presidents may also be a source of influence since their support can add legitimacy to particular policy issues, but increasingly their influence is spread among a growing number of priorities.[4]

The problem with lobbying Bank boards, however, is the diffuse governance structures of the three MDBs. It is difficult to imagine, for example, that strong lobbying of the EIB's member state board members would translate easily into stronger environmental policies at the EIB. This is because the Bank has a nonresident board that consists of minister-level economics officials as well as some private sector members who are not directly accountable to anyone. A better tactic might be to pressure the environmental directorate at the European Commission, since the Commission has veto power on the EIB board. A related lesson is that member states have more leverage in pressuring an MDB to change its policy goals during times of capital replenishment. NGOs who pointed out disastrous World Bank projects to United States Congressional officials in the 1980s gave those officials greater power to insist on policy changes at the Bank during capital replenishment discussions. A number of other important policy changes at the World Bank, and to a lesser extent at the EBRD, also occurred during the process of capital replenishment negotiations.

A second major policy lesson is that there are limits to MDB leverage. First, the World Bank's ability to impose policy conditionality declines as a recipient country gains access to other sources of funding. Second, policy leverage is also a function of how demand-driven an MDB is. If an MDB tends to wait for clients to approach it with partially or fully designed funding proposals, the MDB has less ability to influence a project. Ironically, attempts by MDBs to gain leverage can also backfire and result in more potentially dangerous projects than would otherwise have existed. This is a lesson one can take from the failed Mochovce loan at the EBRD. Pressure for the EBRD to attach conditionality to this loan resulting in a safer nuclear plant was one of the reasons the Slovaks chose to finance the completion of the plant through other sources, which is likely to result in fewer safeguards than would have existed under an EBRD loan.

Finally, while there are potential sources of synergy among the three banks' activities—where they complement each other or work with each other—it is also important to pay attention to examples of where one MDB undercuts another. This has occurred mainly where the EIB has snapped up projects turned down by the other two MDBs, or where the EIB's cheaper prices take business away from the other two. Donors should avoid situations where the EIB is in a position to undermine World Bank conditionality. Recipient governments should not be able to play the banks against each other for the lowest level of conditionality.

It is especially important to strengthen the environmental behavior of the EIB, because ultimately it will be the only one of the three MDBs to remain in CEE. Both the EBRD and World Bank have "graduation" policies, whereby countries reaching a certain level of development are no longer eligible to receive their loans. Indeed, if the EBRD is successful in its work to promote private sector development in CEE and the former Soviet Union, theoretically it should make itself redundant. Its closure would be proof of its success. The presence of the EIB in Central and Eastern European countries preparing to join the EU, by contrast, is growing and will continue to grow.

The world has changed dramatically since the first MDB—the World Bank—was established in 1944 at Bretton Woods as almost an afterthought to the delegates' main goal of creating the International Monetary Fund to help promote a stable global economic order. When Harry Dexter White first wrote a draft proposal for an international bank in early 1942, he made no mention of development, but instead emphasized the importance of a new institution to supply the huge amount of capital needed for postwar reconstruction and economic recovery.[5] The roles of the World Bank and other MDBs have evolved considerably since those days, as conceptions of what "development" means continues to expand and encompass an increasing array of meanings. The shifting definitions are taking place in a rapidly changing international context, where globalization has presented these institutions with new tasks and challenges. The fundamental conflict between the MDB as a financial institution and a development agency is now being stretched even further through moves by international financial institutions to become involved in issues like good governance, institution building, and human rights. The growing number of mandates can cause an MDB to be the victim of its own success, if they contribute to poor or uneven performance. Donor governments, the institutions themselves, and members of civil society "watchdogs" need to be more conscious of the consequences of their efforts to reform MDBs. Ironically, many call for the World Bank to have fewer objectives, which essentially would make it more like the EIB, while others call for the EIB to take on more nonfinancial objectives, or to be more like the World Bank. Some of the more extreme criticisms call for fundamentally altering the basic character of the institutions, such as the U.S. congressional commission (The Meltzer Commission) that recommended the World Bank narrow its focus to a limited set of objectives by essentially ending its life as a financial institution, changing its name to "World Development Agency" and giving out grants on issues such as the treatment of tropical diseases.[6] World Bank president James Wolfensohn captured part of the Bank's dilemma succinctly when he mused, "NGOs criticize me for in fact not going far enough. But when I introduce the things on justice, legal systems and corruption, then the criticism is you're trying to do too much."[7]

This study has sought to contribute to understanding the sources of the gaps between policy ideas and performance within a powerful set of international financial institutions. The arguments advanced in this book not only suggest that it is critical for the institutions to better balance their banking and development goals without tipping too far in one direction or the other, but also explain why it is politically and institutionally difficult for them to do so. In the case of the environment, the optimal institutional balance may be where strong due diligence procedures better permeate all of the bank's projects, while dedicated projects exist in areas that dovetail with the banks' existing expertise. Integrating the environment in this way requires a tighter alignment of policy goals and internal incentive systems and a deeper penetration of green bankers throughout the institutions.

Notes

Chapter One: Introduction and Overview

1. World Bank, "Environment Matters: Annual Review" (Washington, D.C.: World Bank, 1996), 1. Strong was Secretary-General, 1992 Rio Earth Summit, and made this statement when he was Senior Advisor to the World Bank President.

2. Lisa Jordan, "Sustainable Rhetoric vs. Sustainable Development: The Retreat from Sustainability in World Bank Development Policy" (Washington, D.C.: Bank Information Center, 1997), 16. Jordan was director of the Washington, D.C.–based NGO, the Bank Information Center, at the time.

3. Gillian Handyside, "EU Hopefuls Face $130 Billion Environmental Bill," *Reuters,* 19 September 1997. The Czech Republic, Estonia, Hungary, Poland, and Slovenia began negotiations with the EU in March 1998. Also hoping to join the EU are Bulgaria, Latvia, Lithuania, Romania, and Slovakia.

4. I use European Union (EU) to refer to the organization both in general and to specific references since 1993. European Community (EC) is used to refer to the organization prior to 1993.

5. The World Bank has also been responsible for carrying out an additional $89 million in projects through the Global Environmental Facility, which involves the Bank, UNDP, and UNEP. It operates with grant financing and is not directly comparable to work undertaken by the other two MDBs.

6. These are discussed in chapter 2. For critical accounts, see Bruce Rich, *Mortgaging the Earth: The World Bank, Environmental Impoverishment, and the Crisis of Development* (Boston: Beacon Press, 1994); and Catherine Caulfield, *Masters of Illusion: The World Bank and the Poverty of Nations* (New York: Henry Holt, 1996).

7. World Bank, "Toward an Environmental Strategy for the World Bank Group," Progress Report/Discussion Draft, April 2000, 3.

8. In the summer of 1944, when the World Bank was conceived, the Commission II (International Bank for Reconstruction and Development) wrote to the Executive Plenary Session (United Nations Monetary and Financial Conference): "The creation of the Bank was an entirely new venture. . . . So novel was it, that no

name could be found for it. Insofar as we can talk of capital subscriptions, loans, guarantees, issue of bonds, the new financial institutions may have some apparent claim to the name of Bank. But the type of shareholders, the exclusion of all deposits and short-term loans, the non-profit basis, are quite foreign to the accepted nature of a Bank. However, it was accidentally born with the name Bank, and Bank it remains, mainly because no satisfactory name could be found in the dictionary for this unprecedented institution." World Bank, *Annual Report* (Washington, D.C., World Bank: 1994), 12.

9. In fact, there are many instances where the banks find themselves competing with private sector lenders. In December 1995, for example, the World Bank's private lending arm, the International Finance Corporation (IFC), set up new guidelines to prevent it from competing with private sector banks in privatization and securities underwriting. See George Graham, "IFC to leave Privatization to Banks," *Financial Times,* 8 December 1995, 4.

10. The EIB, in many ways the most banklike of the three MDBs studied here, also happens to be the most risk-averse. In its work in many parts of the world, it uses "double guarantees," which means its loans can be backed by a government as well as the European Commission.

11. For neorealist/neoliberal debates, see Robert Keohane, *After Hegemony: Cooperation and Discord in the World Political Economy* (Princeton: Princeton University Press, 1984) and John J. Mearsheimer, "The False Promise of International Institutions," *International Security* 19:3 (winter 1994/1995): 5–49. On institutional learning, see Ernst B. Haas, *When Knowledge Is Power: Three Models of Change in International Organization* (Berkeley: University of California Press, 1990) and Peter M. Haas and Ernst B. Haas, "Learning to Learn: Improving International Governance," *Global Governance* 1 (September–December 1995): 255–285. A constructivist argument seeking to explain sources of institutional pathologies can be found in Michael N. Barnett and Martha Finnemore, "The Politics, Power and Pathologies of International Organizations," *International Organization* 53:4 (autumn 1999): 699–732.

12. EIB lending surpassed World Bank lending in 1992 and 1994–1999. In 1999, for example, the EIB loan commitments were 31.8 billion euro ($34 billion, using the end fourth quarter exchange rate of 1 euro = $1.07), versus World Bank commitments of $28.9 billion. World Bank and EIB annual reports, 1992–1999.

13. For early accounts of the region's environmental context, see Richard Ackermann, "Environment in East Central Europe: Despair or Hope," *Transition* 4 (1991): 9–11; Duncan Fisher, "Paradise Deferred: Environmental Policymaking in Central and East Central Europe" (London: Royal Institute of International Affairs and Ecological Studies Institute, 1992); Bedrich Moldan and Jerald L. Schnoor, "Czechoslovakia: Examining a Critically Ill Environment," *Environment Science and Technology* 26:1 (1992); Joan DeBardeleben, ed., *To Breathe Free: East Central Europe's Environmental Crisis* (Baltimore: Johns Hopkins Press, 1991); Hilary F. French, "Eastern Europe's Clean Break with the Past," *Worldwatch* (March/April 1992); Joseph Alcamo, ed., *Coping with Crisis in Eastern Europe's Environment* (New York: Parthenon Publishing Group, 1992);

and Helmut Schreiber, "The Threat From Environmental Destruction in Eastern Europe," *Journal of International Affairs* 44 (winter 1990): 359–391.

14. I am referring to the World Bank's Polonoereste project in Brazil in the 1980s, the Sardar Sarovar dam project in India in the early 1990s, and the EBRD's proposed loans for the Mochovce nuclear plant in Slovakia and the Rivne and Khmelnytsky nuclear plants in the Ukraine in the mid- and late-1990s, respectively. See chapter 2 for discussion of major problem projects.

15. Ibrahim Shihata, *The World Bank Inspection Panel* (New York: Oxford University Press, 1994), 139.

16. World Bank, "The World Bank and the Environment in Central and Eastern Europe: 1990–95" (Washington, D.C.: The World Bank, 1995), 5.

17. Private capital flows to developing countries increased dramatically in the 1990s, even including subdued levels in 1998 and 1999 as a result of the Asian and Russian financial crises of 1997–1998. Net private capital flows to twenty-nine major developing countries stood at $150 billion in 1999. The vast majority of these flows to developing countries is accounted for by a dozen countries, which means that MDBs remain significant actors in most parts of the world. Nonetheless, the surge in private flows has prompted MDBs to develop new strategies for working with the private sector. See Institute for International Finance, "Capital Flows to Emerging Market Economies" (Washington, D.C., September 25, 1999), at ⟨http://www.iif.com/Press/Rel/1999pr13.html⟩.

18. Some MDBs, such as the World Bank, have taken on a variety of other issues over the years, ranging from judicial reform to the promotion of gender equality. Others, such as the EIB, have not. Yet, even for the World Bank, the environmental mandate has received the most attention and the most emphasis in the Bank's budget.

19. For an example of work praising the Bretton Woods institutions, see Henry Owen, "The World Bank: Is 50 Years Enough?" *Foreign Affairs* 73, 5 (September/October 1994), 97–108. For an example of work arguing for a vastly more limited World Bank, or even the closure of the Bank, see Kevin Danaher, ed., *Fifty Years Is Enough* (Boston: South End Press, 1994). In the center are other attempts to propose reforms, such as the Bretton Woods Commission Report, which recommended ways in which the IMF and the World Bank can refocus their energies; see Bretton Woods Commission, *Bretton Woods: Looking to the Future* (Washington, D.C.: July 1994). Slightly more critical and focusing solely on the Bank is Peter Brosshard, et al., *Lending Credibility: New Mandates and Partnerships for the World Bank* (Washington, D.C.: World Wildlife Fund, 1996).

20. Donald M. Goldberg, et al., "The European Bank for Reconstruction and Development: An Environmental Progress Report" (Washington, D.C.: CIEL, 1995).

21. World Commission on Environment and Development, *Our Common Future* (New York: Oxford University Press, 1987), 21.

22. The UNCED secretariat estimated the cost of implementing Agenda 21, the guiding document produced at UNCED, to be around $125 billion a year. See

Michael Grubb et al., *The "Earth Summit" Agreements: A Guide and Assessment* (London: Earthscan Publications, 1993).

23. Haas and Haas, "Learning to Learn," 266. The other institution is UNEP. Haas and Haas examined a set of thirteen IOs, of which the World Bank was the only MDB.

24. Rich, *Mortgaging the Earth,* 281.

25. Friends of the Earth, *Testimony before Congress on World Bank Appropriations,* 1 March 1993, 14, cited in Caulfield, *Masters of Illusion,* 268.

26. The "World Bank" refers to the International Bank for Reconstruction and Development (IBRD) and the International Development Association (IDA). IDA is an affiliate that provides interest-free loans to the poorest countries; in CEE, Albania is the only eligible country. In addition, the bank has three other affiliates: the Multilateral Investment Guarantee Agency (MIGA), which insures private sector investors against noncommercial (political) risks; the International Finance Corporation (IFC), which is the World Bank's private sector lending arm; and the International Centre for Settlement of Investment Disputes (ICSID), which offers conciliation and arbitration facilities for investment disputes between foreign investors and recipient governments. The IBRD, IDA, MIGA, IFC, and ICSID together are referred to as the World Bank Group.

27. As noted in chapter 5, the largest single donor to the region is Germany, followed by the World Bank, the EIB, EU's PHARE program, the EBRD, and the U.S. Support for East European Democracy program.

28. For example, net private capital flows to emerging market economies in Europe stood at $37.6 billion in 1999, compared with net external financing from international financial institutions and bilateral creditors of $0.3 billion, according to Institute for International Finance data. Countries included in this data are Bulgaria, the Czech Republic, Hungary, Poland, Romania, Russia, Slovakia, and Turkey. See Institute for International Finance, "Capital Flows to Emerging Market Economies" (Washington, D.C., 31 May 2001), 4.

29. For the most thorough description and analysis of the region's environmental problems, see the World Bank–led report: "Environmental Action Programme for Central and Eastern Europe," document submitted to the Ministerial Conference in Lucerne, Switzerland (28–30 April, 1993).

Chapter Two: Intellectual Context: Understanding MDB Greenness

1. World Bank, *Environment Matters,* fall 1996, 5.

2. Barbara Connolly and Tamar Gutner, "Policy Networks and Process Diffusion: Environment for Europe" (paper presented at the annual meeting of the International Studies Association conference, Los Angeles, Calif., March 2000).

3. According to a draft report by the World Bank's Quality Assurance Group, poor project performance can often be traced to "inadequate project design," and "the most common shortcoming is excessive complexity." World Bank, memorandum, *Portfolio Improvement Program: Draft Reviews of Sector Portfolio and Lending Instruments: A Synthesis,* 23 April 1997, 12.

4. Some MDB projects do propose indicators for environmental impact. For example, some of the World Bank's wastewater projects in the Baltic Sea region do have water quality and environmental indicators, including those for nitrogen removal, phosphorus removal, and BOD7. *Water Quality and Environmental Indicators*, ⟨http://water.hut/fi/BUBI/home/benchmarking/environmental.html⟩.

5. Kerstin Canby, World Bank Environmental Department, interview with the author, March 1998.

6. "Integrated resource planning" is the approach by which decisions about utility investments take both supply- and demand-side energy conservation options into account. Michael Philips, *The Least-Cost Energy Path for Developing Countries* (Washington, D.C.: International Institute for Energy Conservation, 1991).

7. EIB, *Environmental Policy Statement* (Luxembourg: European Investment Bank, 1996).

8. World Bank, *Making Development Sustainable* (Washington, D.C.: World Bank, 1994), 28.

9. Pollution management and urban environmental projects address issues such as improving solid waste management and rehabilitating municipal water treatment; rural environmental projects address issues such as forestry management and watershed rehabilitation; and institutions projects seek to build up the management capacity of national and local environmental agencies. World Bank, *Mainstreaming the Environment* (Washington, D.C.: World Bank, 1995), 4–6.

10. Environmental Defense Fund, "World Bank Plans Increased Support for Timber Harvesting," 16 May 1994, http://www.edf.org/pubs/newsreleases/1994/may/e%Fwforestr.html. See also, Caulfield, *Masters of Illusion*, 186.

11. World Bank, *Annual Report* (Washington, D.C.: World Bank, 1994), 102.

12. World Bank, *Making Development Sustainable*, 217–218. One project was a $38.4 million district heating rehabilitation project in Estonia, which was listed as an oil and gas project in the annual report; the other was a $146 million forest development project in Poland, which was not clearly categorized in the annual report.

13. World Bank, "Central Europe Department Projects Related to Energy/Environment" (World Bank, Washington, D.C., August/September 1992, photocopy).

14. World Bank, *The World Bank and the Environment in Central and Eastern Europe: 1990–95* (Washington, D.C.: World Bank, 1995), 19.

15. Figures for fiscal 1990–2000 in World Bank, "Annual Report: Financial Statements, Appendixes, Project Summaries" (Washington, D.C.: World Bank, 2000), 143.

16. Joan M. Nelson and Stephanie J. Eglinton, *Global Goals, Contentious Means: Issues of Multiple Aid Conditionality* (Washington, D.C.: Overseas Development Council, 1993).

17. For a rationalist framework, see Andrew Moravcsik, *The Choice for Europe: Social Purpose and State Power from Messina to Maastricht* (Ithaca: Cornell University Press, 1998); on unintended consequences, see Paul Pierson, "The Path

to European Integration," *Comparative Political Studies* 29, 2 (April 1996): 123–163; on constructivist approaches focusing on the entry of ideas into international institutions, see Haas and Haas, "Learning to Learn."

18. See, for example, Rich, *Mortgaging the Earth;* Philippe G. Le Prestre, *The World Bank and the Environmental Challenge* (Cranbury, N.J.: Associated University Presses, 1989); and Robert Wade, "Greening the Bank," in *The World Bank: Its First Half Century,* eds. Devesh Kapur, John P. Lewis, and Richard Webb, vol. 2 (Washington, D.C.: The Brookings Institution, 1997), 611–736. Caufield, *Masters of Illusion: The World Bank and the Poverty of Nations,* contains many sections on the Bank's environmental troubles in her broader critique of the World Bank.

19. On the EBRD, see Goldberg et al., "The European Bank for Reconstruction and Development: An Environmental Progress Report"; and Chris A. Wold and Durwood Zaelke, "Promoting Sustainable Development and Democracy in Central and Eastern Europe: The Role of the European Bank for Reconstruction and Development," *American University Journal of International Law and Policy* (1992): 559–564. On the EIB, see Marianne Wenning, "Greening the European Investment Bank" (Switzerland: WWF International, December 1992); Sheila Lewenhak, *The Role of the European Investment Bank* (London: Croom Helm, 1982); and Carl Lankowski, "Environmental Impact Review in the European Investment Bank" (Washington, D.C.: School of International Service, American University, May 1984).

20. The Polonoroeste and Sardar Sarovar projects are described below. For details on a number of these failed projects, see Rich, *Mortgaging the Earth.*

21. Graham Bowley, "MDBs under Investigation," *Financial Times,* 27 September 1996, xxi.

22. They define "problematique" as "a given set of interdependent problems, places, acts, and policies—such as those associated with sustainable development." Haas and Haas, "Learning to Learn," 257. The EBRD and EIB were not among the institutions included in this study.

23. Rich, *Mortgaging the Earth.*

24. Rich writes, "Polonoroeste transformed Rondônia—an area approximately the size of Oregon or Great Britain—into a region with one of the highest rates of forest destruction in the Brazilian Amazon, increasing its deforested area from 1.7 percent in 1978 to 16.1 percent in 1991. By the mid-1980s, the burning of Rondônia's forests became a major focus of NASA research as the single largest, most rapid human-caused change on earth readily visible from space." Ibid., p. 28. Also see Wade, "Greening the Bank."

25. The Sardar Sarovar project, one of the largest water resources projects ever planned, involved the construction of a dam, major canal, and irrigation network along India's Narmada River. The project's goals were to provide more irrigable land, to bring more drinking water to a drought-prone region, and to create a major source of power. At least 100,000 people lived in the villages that would be submerged. Environmentalists argued that the affected populations were not properly consulted and would not be properly resettled, that the Bank supported

the project even when it became clear that the borrower was not meeting the loan conditions, and that the environmental impacts of the project were not properly addressed. Bradford Morse and Thomas Berger, *Sardar Sarovar: Report of the Independent Review* (Ottawa: Resources Future International, 1992), 4. (This was also known as the Morse Commission Report.)

26. Rich, *Mortgaging the Earth,* 250.

27. Morse and Berger, *Sardar Sarovar,* xx.

28. Ibid., xxi.

29. Ibid., 226. Despite the report's recommendation that the Bank withdraw from the project, and despite the fact that the United States, Germany, Japan, Canada, Australia and the Nordic countries (with 42 percent of the vote) were in favor of suspension, the board of directors voted to continue. See Wade, "Greening the Bank," 704. A year later, when it looked like a majority of the board favored canceling the loan, the government of India decided not to ask for additional Bank funding of the project, which Rich explains was a face-saving effort for both the Indian government and the Bank. See Rich, *Mortgaging the Earth,* 302.

30. The Arun III dam project was criticized by NGOs as being too costly, thus using funding that might be better allocated elsewhere; for being too big, when a smaller, cheaper dam might be more appropriate; for producing much more electricity than the country needed, meaning that much of it would be exported to India; and for not contributing to poverty alleviation in the country. Thus, while environmental groups again led criticism of the project, their complaints were not directly tied to environmental issues. For one view on the Arun case, see Bruce Rich, "Epilogue: The Gorbachav of the World Bank?" (Environmental Defense Fund, Washington, D.C., 1997, photocopy).

31. For analysis of the mixed outcomes of the World Bank's forestry policies, see Michael Ross, "Conditionality and Logging Reform in the Tropics," in *Institutions for Environmental Aid: Pitfalls and Promise,* ed. Robert O. Keohane and Marc A. Levy (Cambridge: The MIT Press, 1996), 167–197. On energy policy, see *Power Failure: A Review of the World Bank's Implementation of Its New Energy Policy* (Washington, D.C.: Environmental Defense Fund, Natural Resources Defense Council, March 1994). The EDF/NRDC report, for example, reviewed all of the Bank's power loans that were being prepared during the first half of 1993 and compared them to the Bank's new energy policies. The majority of the Bank's power loans were found to be weak in their compliance to the Bank's policies.

32. *Effective Implementation: Key to Development Impact.* Portfolio Management Task Force. Confidential Report of the World Bank (Washington, D.C., September 22, 1991), ii.

33. Ibid.

34. *Portfolio Improvement Program,* 1.

35. David Reed, ed., *Structural Adjustment and the Environment* (Boulder: Westview Press, 1992).

36. Berg and Sherk, for example, point out examples where export crops result in less soil erosion than food crops. See Elliot Berg and Don Sherk, "The World Bank and Its Environmentalist Critics," in *Bretton Woods: Looking to the Future* (Washington, D.C.: Bretton Woods Commission, July 1994), C-316. The World Bank's own work on SALs and the environment has concluded that SALs produce positive environmental benefits where environmental policies are already in place. Otherwise, the results were mixed. Currency devaluations, for example, generally raise farmgate prices, but the impact on the environment depended on which crops were encouraged as a result. See David W. Pearce and Jeremy J. Warford, *World without End: Economics, Environment, and Sustainable Development* (New York: Oxford University Press, 1993), 316.

37. The Bank, meanwhile, has generated quite a library of publications to inform the public on its growing interest in and attention to environmental issues. It now has annual reports on its environmental activities, as well as an annual conference on environmentally sustainable development.

38. See Wade, "Greening the Bank."

39. World Bank, Operations Evaluation Department, Sector and Thematic Evaluation Group, "Promoting Environmental Sustainability in Development: An Evaluation of the World Bank's Performance," September 15, 2000.

40. Ibid.

41. Marianne Wenning, "Greening the European Investment Bank," WWF International Discussion Paper, December 1992.

42. CEE Bankwatch Network, "The European Investment Bank: Accountable Only to the Market?" EU-Policy Paper no. 1, Brussels: Heinrich Böll Foundation, 1999, 3.

43. Öko-Institute, "Statement Concerning the Least Cost Study for the Public Participation Programme Related to the Project 'Completion of the Mochovce NPP (Slovak Republic)'" (Freiburg, Germany: Institute für Angewandte Ökologie, 1995). Curiously, one critic of this study was a sister development bank of the EBRD, the EIB, which not only shares most of the same major donors as the EBRD but is itself a shareholder of the EBRD.

44. "NGOs Criticise EBRD Lending, Environment Record," *Reuters News Service,* 23 April 2001.

45. Moises Naim, "The World Bank: Its Role, Governance, and Organizational Culture," in *Bretton Woods: Looking to the Future,* C-276.

46. Payer argues that the United States has "always been able to control the direction" of World Bank lending, and cites numerous examples of the United States imposing its wishes on Bank lending. See Cheryl Payer, *The World Bank: A Critical Analysis* (New York: Monthly Review Press, 1982).

47. Rich, *Mortgaging the Earth,* 305.

48. Naim, "The World Bank: Its Role, Governance, and Organizational Culture," C-281.

49. Inspired by Greenpeace, "World Bankenstein: A monstrous institution," *Greenpeace* 3 (4) (October–December 1994), 2.

50. George and Sabelli, *Faith and Credit.*

51. This is cited in U.S. Congress, Senate, Subcommittee on International Economic Policy, Trade, Oceans and Environment, Committee on Foreign Relations, *Statement of Bruce M. Rich on Behalf of Environmental Defense Fund, Friends of the Earth, National Audubon Society, National Wildlife Federation, Sierra Club Concerning Public International Financial Institutions: Environmental Performance and Management,* 3 March 1994. Rich attributes the quote to minutes from oral statements of the World Bank's Executive Directors, held 23 October 1992.

52. D. Roderick Kiewiet and Mathew D. McCubbins, *The Logic of Delegation: Congressional Parties and the Appropriations Process* (Chicago: University of Chicago Press, 1991), chapter 2.

53. Robert W. Cox and Harold K. Jacobson, *The Anatomy of Influence: Decision Making in International Organization* (New Haven: Yale University Press, 1973), 7. On historical institutionalism and other institutionalisms, see Peter A. Hall and Rosemary C. R. Taylor, "Political Science and the Three Institutionalisms," *Political Studies* 44 (1996): 952–973.

54. Alexander Wendt, "Anarchy Is What States Make of It: the Social Construction of Power Politics," *International Organization* 46 (2) (1992): 391–425. Barnett and Finnemore draw on Weber's classic ideas of bureaucratic behavior to argue that IO bureaucracies have sources of autonomy based on their embodiment of "rational-legal authority" and their ability to control technical information and expertise. Barnett and Finnemore, "The Politics, Power, and Pathologies of International Organizations."

55. See discussion in Mark A. Pollack, "Delegation, Agency, and Agenda Setting in the European Community," *International Organization* 51 (1) (1997), 109. These ideas are also linked to the arguments of students of bureaucratic politics, such as Max Weber.

56. Kathleen Thelen and Sven Steinmo, "Historical Institutionalism in Comparative Politics," in *Structuring Politics: Historical Institutionalism in Comparative Analysis,* eds. Sven Steinmo, Kathleen Thelen, and Frank Longstreth (New York: Cambridge University Press, 1992), 3.

57. Kiewiet and McCubbins, *The Logic of Delegation,* chapter 2.

58. Susan Strange, "Cave! hic dragones: A Critique of Regime Analysis," in *International Regimes,* ed. Stephen Krasner (Ithaca: Cornell University Press, 1983), 337–354; see also Joseph Grieco, "Anarchy and the Limits of Cooperation: A Realist Critique of the Newest Liberal Institutionalism," *International Organization* 42 (3) (1988): 485–507.

59. Mearsheimer, "The False Promise of International Institutions," 7.

60. Robert O. Keohane, *After Hegemony: Cooperation and Discord in the World Political Economy* (Princeton: Princeton University Press, 1984).

61. On the importance of institutional form, see John Gerard Ruggie, "Multilateralism: The Anatomy of an Institution," in *Multilateralism Matters,* ed. John Gerard Ruggie (New York: Columbia University Press, 1993), 3–47. Also in this category are transaction cost theories of organizational structure. See Oliver

Williamson, "The Economics of Organization: The Transaction Cost Approach," *American Journal of Sociology* 87 (3) (1981): 548–577.

62. Lisa L. Martin and Beth A. Simmons, "Theories and Empirical Studies of International Institutions," *International Organization* 52 (4) (autumn 1998): 729–757.

63. By contrast, at the other two banks the richer shareholder countries, which provide the bulk of the banks' capital, have more say on the conditionality of loans going to poorer recipients, which are also shareholders.

64. I refer here to the Environmental Action Programme for Central and Eastern Europe, mainly designed by the World Bank. It is discussed in chapter 5.

65. This is called the "negative transfer" problem. See Rich, *Mortgaging the Earth,* 183.

66. This point is reiterated by Horberry in a study of factors shaping the accountability of development agency staff. Horberry argued that World Bank staff will resist new policy demands, such as environmental mandates, if they conflict with internal institutional interests, such as the need to move money. John Horberry, "The Accountability of Development Assistance Agencies: The Case of Environmental Policy," *Ecology Law Quarterly* 12 (1985), 824.

67. Wapenhans Report, iii.

68. Frances F. Korten, "The High Costs of Environmental Loans," *Asia Pacific Issues: Analysis from the East-West Center* 9 (1993): 1–8.

69. Wade, "Greening the Bank."

70. For a succinct discussion of preference-driven models versus institutional-driven models in explaining public policy formation, see Peter Alexis Gourevitch, "Squaring the Circle: The Domestic Sources of International Cooperation," *International Organization* 50 (2) (spring 1996): 349–373.

71. See, for example, Walter Mattli and Anne-Marie Slaughter, "Revisiting the European Court of Justice," *International Organization* 52 (1) (winter 1998): 177–209.

72. Pierson, "The Path to European Integration."

73. Pollack, "Delegation, Agency, and Agenda Setting in the European Community."

74. The work on EU integration tends to focus on one EU institution, such as the ECJ, or addresses the different institutions, including the ECJ, Commission, or Council. Curiously, none of this work has included the EIB.

75. Kenneth A. Shepsle, "Studying Institutions: Some Lessons from the Rational Choice Approach," *Journal of Theoretical Politics* 1 (2) (1989): 131–147. The public choice movement overall emphasizes heavily how institutional arrangements explain collective choice. See James Buchanan and Gordon Tullock, *The Calculus of Consent* (Ann Arbor: University of Michigan Press, 1962).

76. The most relevant sociological organization theory works include James G. March and Herbert A. Simon, *Organizations* (New York: John Wiley and Sons, 1958); James G. March and Johan P. Olsen, *Rediscovering Institutions: The Or-*

ganizational Basis of Politics (New York: The Free Press, 1989); and Walter W. Powell, Paul J. DiMaggio, eds., *The New Institutionalism in Organizational Analysis* (Chicago: University of Chicago Press, 1991). On bureaucratic politics, see Anthony Downs, *An Economic Theory of Democracy* (New York: Harper and Row, 1957). On the importance of decision-making procedures, see Graham Allison, *Essence of Decision: Explaining the Cuban Missile Crisis* (Boston: Little, Brown, 1971).

77. James G. March, "Bounded Rationality, Ambiguity, and the Engineering of Choice," *Bell Journal of Economics* 9 (2) (1978): 587–608.

78. Oran Young, *International Governance: Protecting the Environment in a Stateless Society* (Ithaca: Cornell University Press, 1994), 154.

79. Barbara Connolly, Tamar Gutner, and Hildegard Bedarff, "Organizational Inertia and Environmental Assistance to Eastern Europe," in *Institutions for Environmental Aid,* 282. On the garbage can model, see Michael D. Cohen, James G. March, et al., "A Garbage Can Model of Organizational Choice," *Administrative Science Quarterly* 17 (1972): 1–25.

80. Pat Aufderheide and Bruce Rich, "Environmental Reform and the Multilateral Development Banks," *World Policy Journal* 5 (spring 1988): 306.

81. George and Sabelli, *Faith and Credit,* 169.

82. Cited in Nelson and Eglinton, *Global Goals, Contentious Means,* 24.

83. Barbara Connolly, "Increments for the Earth: The Politics of Environmental Aid," in *Institutions for Environmental Aid,* 328. For an excellent analysis of the ways in which lessons from the economic development literature are salient to environmental aid, see David Fairman and Michael Ross, "Old Fads, New Lessons: Learning from Economic Development Assistance," in *Institutions for Environmental Aid,* 29–51.

84. Paul Mosley, Jane Harrigan, and John Toye, *Aid and Power: The World Bank and Policy-based Lending,* vol. 1 (New York: Routledge, 1991), 69.

85. Ross, "Conditionality and Logging Reform in the Tropics," 189.

86. For an excellent explication on different forms of capacity, see Connolly, "Increments for the Earth: The Politics of Environmental Aid," 345.

87. Ibid., 328.

88. Naim, "The World Bank: Its Role, Governance, and Organizational Culture," C-276.

89. Ibid., C-273.

90. Horberry, "The Accountability of Development Assistance Agencies," 818, 833. For other versions of this view, see Eugene H. Rotberg, "Financial Operations of the World Bank," in *Bretton Woods: Looking to the Future,* C-185–214; and Judith Tendler, *Inside Foreign Aid* (Baltimore: Johns Hopkins University Press, 1975).

91. William Ascher, "New Development Approaches and the Adaptability of International Agencies: The Case of the World Bank," *International Organization* 37 (3) (summer 1983), 437.

Chapter 3: Bargaining and Delegation: The Birth of Environmental Mandates

1. EBRD, "Agreement Establishing the European Bank for Reconstruction and Development," Art. 2 (1)(vii).

2. EIB, *Environmental Policy Statement*, 1996, 4.

3. John W. Kingdon, *Agendas, Alternatives, and Public Policies* (Boston: Little, Brown, 1984), particularly chapter 8. The model draws from the Cohen, March, and Olsen "garbage can" model of organizational decisionmaking, which has four streams (problems, solutions, participants, and choice opportunities), compared to Kingdon's three streams. Kingdon's model also emphasizes patterns and structure in the linking of these processes, whereas the earlier model takes a more anarchical view. On the garbage can model, see Michael D. Cohen, James G. March, and Johan P. Olsen, "A Garbage Can Model of Organizational Choice," *Administrative Science Quarterly* 17 (1972): 1–25.

4. In addition to the EBRD and EIB, the World Bank's sister regional MDBs include the Inter American Development Bank, established in 1959; the African Development Bank, established in 1964; and the Asian Development Bank, established in 1966.

5. For histories of the World Bank, see Kapur, Lewis, and Webb, eds., *The World Bank: Its First Half Century*, 2 vols.; Edward S. Mason and Robert E. Asher, *The World Bank Since Bretton Woods* (Washington, D.C.: The Brookings Institution, 1973); Catherine Gwin, *U.S. Relations with the World Bank* (Washington, D.C.: The Brookings Institution, 1994); Jochen Kraske et al., *Bankers with a Mission: The Presidents of the World Bank, 1946–91* (Washington, D.C.: Oxford University Press, 1996); Susan George and Fabrizio Sabelli, *Faith and Credit: The World Bank's Secular Empire* (Boulder: Westview Press, 1994); and Rich, *Mortgaging the Earth*.

6. Gwin, *United States Relations with the World Bank*, 3.

7. World Bank Articles of Agreement, in Mason and Asher, *The World Bank Since Bretton Woods*, 779.

8. World Bank, Annual Report (Washington, D.C.: World Bank, 1999), 278.

9. Article III, Section 5 (b).

10. Nonetheless, the Bank's first four loans were to France, the Netherlands, Denmark, and Luxembourg, for reconstruction purposes.

11. Mason and Asher, *The World Bank Since Bretton Woods*, 467.

12. For an excellent and critical account of the Bank's interactions with and contributions to the field of development economics, see Nicholas Stern and Francisco Ferreira, "The World Bank as 'Intellectual Actor,'" in *The World Bank: Its First Half Century*, vol. 2, 523–610.

13. During McNamara's tenure, the Bank's staff size more than tripled, from 1,600 to 5,700, and its lending shot up to over $13 billion from around $1 billion. Kraske et al., *Bankers with a Mission*, 179, 213. See Rich, *Mortgaging the Earth*

for a particularly scathing critique of the McNamara era, which he sees as resulting in an institution "where policy pronouncements and rhetoric were largely dissociated from reality" (102).

14. While this book does not analyze the Bank's performance on issues outside the environment, there is no shortage of criticism on the Bank's efforts to address other issue areas. See Michelle Miller-Adams, *The World Bank: New Agendas in a Changing World* (New York: Routledge, 1999) for a comparative analysis of the Bank's mixed efforts in three areas: private sector development, public participation, and governance.

15. Stern and Ferreira, "The World Bank as 'Intellectual Actor,'" 523. Indeed, the Bank is an important source of economic data on the developing world.

16. This is the World Bank's figure and includes active and closed projects. Stand-alone projects are defined as including natural resource management, pollution and waste management, and institutional development projects. World Bank, "Environment Matters" (Washington, D.C.: World Bank, 2001), 56–57.

17. In defining the Bank's goals for the decade, McNamara called for continued attention to be given to issues such as population planning, education, and agriculture, and argued for a more active Bank role in addressing the interrelated problems of unemployment, urbanization, and industrialization. See Robert McNamara, "To the Board of Governors, Washington, D.C., 29 September 1969," in *The McNamara Years at the World Bank: Major Policy Addresses of Robert S. McNamara 1968–1981* (Baltimore: Johns Hopkins University Press, 1981), 69–74. His comments on the environment were made during a speech at a session of the United Nations Economic and Social Council. Cited in Le Prestre, *The World Bank and the Environmental Challenge*, 19.

18. By the early 1980s, the name had changed once more, to the Office of Environmental and Scientific Affairs.

19. For most developing countries at the time, economic growth was the overwhelming priority and environmental worries were thought to be a rich country problem. For more in-depth analyses of the rise of environmentalism and global environmental concerns, see Lynton Keith Caldwell, *International Environmental Policy: Emergence and Dimensions*, 2nd rev. ed. (Durham, N.C.: Duke University Press, 1990), and Tony Brenton, *The Greening of Machiavelli* (London: The Royal Institute of International Affairs, 1994).

20. The Soviets and Eastern Europeans dropped out of the preparatory process and the Stockholm Conference, with a view that pollution was a problem of capitalist countries.

21. Wade, "Greening the Bank," 623.

22. Le Prestre, *The World Bank and the Environmental Challenge*, 23.

23. Brenton, *The Greening of Machiavelli*, 36. See Barbara Ward and Rene Dubos, *Only One Earth: The Care and Maintenance of a Small Planet* (New York: Norton, 1972).

24. See Serge Taylor, *Making Bureaucracies Think: The Environmental Impact Statement Strategy of Administrative Reform* (Stanford: University of California Press, 1984).

25. According to Le Prestre, West Germany and Japan were "reticent," while France was "largely indifferent." Le Prestre, *The World Bank and the Environmental Challenge*, 21.

26. The Treasury, in particular, was opposed to the sharp growth in World Bank lending during the '70s, while McNamara believed that the decline in economic growth resulting from higher oil prices necessitated greater lending. See Gwin, *U.S. Relations with the World Bank*, 20–25.

27. Wade, "Greening the Bank," 635.

28. Le Prestre, *The World Bank and the Environmental Challenge*, 26.

29. Lee headed the environmental office from its creation until 1987. Wade, "Greening the Bank," 620.

30. Le Prestre, *The World Bank and the Environmental Challenge*, 31.

31. Wade, "Greening the Bank," 621.

32. Barber B. Conable, "To the World Resources Institute, Washington, D.C., May 5, 1987," in *The Conable Years at the World Bank: Major Policy Addresses of Barber B. Conable, 1986–91* (Washington, D.C.: The World Bank, 1991), 22.

33. Ibid., 23.

34. Wade, "Greening the Bank," 674–675.

35. Staff in the Bank's environmental department, for example, have included people who work on financial, procurement, and business analysis. It is also difficult to know when to classify sociologists, dam engineers, or fisheries' economists as "environmental" or not.

36. See the description of the Polonoroeste project in chapter 2. The Indonesian Transmigration project was criticized as a disaster from environmental and social grounds, as was Polonoroeste. The goal of the project was to relocate millions of people from Indonesia's inner islands to its less populated outer islands. The project was criticized as being politically motivated. It resulted in the clearing of enormous tracts of tropical forest, which caused rapid deforestation, and did little to alleviate poverty. See Rich, *Mortgaging the Earth*, 34–38.

37. IDA's funds are replenished on a three-year cycle. IBRD capital increases occur less often.

38. Gwin, *United States Relations with the World Bank*, 18–19.

39. Rich, *Mortgaging the Earth*, 112.

40. Wade, "Greening the Bank," 658.

41. Rich, *Mortgaging the Earth*, 138. Incidentally, one of these House subcommittees was chaired by Democrat James Scheuer, who would later become the U.S. Executive Director at the EBRD, where he was a leading actor pushing the Bank to promote energy efficiency.

42. Rich, *Mortgaging the Earth*, 137.

43. Wade, "Greening the Bank," discusses this in greater depth on 668.

44. Rich, *Mortgaging the Earth*, 145.

45. The Fund was established with $160 million in capital but quickly rose to $240 million when India and China joined, during the 1991–1993 period. In 1993, the interim fund was converted to a permanent fund, and its capital was increased to $510 million for the three year period through 1996. For details, see Ibrahim F. I. Shihata, *The World Bank in a Changing World: Selected Essays and Lectures,* vol. 2 (Boston: Martinus Nijhoff Publishers, 1995), 229–230; and DeSombre and Kaufmann, "The Montreal Protocol Multilateral Fund: Partial Success Story," in *Institutions for Environmental Aid,* 89–126.

46. For a description of this campaign, see Ibrahim F. I. Shihata, *The World Bank Inspection Panel* (New York: Oxford University Press, 1994), 22–124.

47. The new policy made more Bank documents available to the public and created a new public information center.

48. Whereas the other two MDBs have specific "environmental policy" documents one can turn to for a summary of their policy goals, the World Bank has produced a plethora of documents, including an annual report on the environment, initiated in the early 1990s, as well as the influential 1992 World Development Report on environment and development.

49. For an excellent analysis of the politics surrounding the creation of the EBRD, see Steven Weber, "Origins of the European Bank for Reconstruction and Development," *International Organization* 48 (1) (winter 1994): 1–38. For an examination of the EBRD's Articles of Agreement vis-à-vis those of other MDBs, see Ibrahim F. I. Shihata, *The European Bank for Reconstruction and Development: A Comparative Analysis of the Constituent Agreement* (London: Graham and Trotman/Martin Nijhoff, 1990).

50. George Graham, "France to Press Proposals for European Development Bank," *Financial Times,* 1 December 1989, 4.

51. Ibid.

52. Tamar Gutner, "Bankrolling Eastern Europe," *Prodigy News Service,* December 1990.

53. Stephan Haggard and Andrew Moravcsik, "The Political Economy of Financial Assistance to Eastern Europe, 1989–1991," in *After the Cold War: International Institutions and State Strategies in Europe, 1989–1991,* eds. Robert O. Keohane, Joseph S. Nye, and Stanley Hoffmann (Cambridge: Harvard University Press, 1993), 246–285.

54. According to journalist George Graham, the view of French officials was that "there can be no question of the Soviet Union . . . sitting on (the EIB) board and deciding on whether to fund projects in, say, Portugal." Graham, "France to Press Proposals for European Development Bank," 4.

55. Ernst-Günther Bröder, interview with the author, 26 March 1997. Bröder was EIB president at the time of the EBRD's birth. He added that the EIB was involved in setting up the EBRD, in areas such as the development of the EBRD's Treasury Department.

56. EBRD, "Agreement Establishing the European Bank for Reconstruction and Development," Art. 1.

57. Ibid., Art. 2 (1)(vii).

58. Of the 2.1 billion ECU in financing approved in 1993, for example, only 409 million ECU was disbursed. News that in 1992 the Bank spent twice as much outfitting its offices as it disbursed in loans did not help its reputation and increased pressure for Attali's resignation. See Robert Preston and Jimmy Burns, "EBRD Spends More on Itself Than It Hands Out in Loans," *Financial Times,* 13 April 1993, 1.

59. Nicholas Brady to William K. Reilly, 22 February 1990. Among other major donor countries, Japan was passive toward the initiative, but not critical of it. See Dennis T. Yasutomo, *The New Multilateralism in Japan's Foreign Policy* (New York: St. Martin's Press, 1995), 173–174.

60. Statement by the Honorable David C. Mulford, Under Secretary of Treasury for International Affairs before the Senate Foreign Relations Committee, Subcommittee on International Economic Policy, Trade, Oceans, and the Environment, 22 March 1990.

61. CEE's environmental plight is discussed in detail in chapter 5. It is important to note that the strong CEE constituencies pushing for environmental reform before the regime changes disappeared quite rapidly when the overwhelming challenges related to economic restructuring made environmental improvement a lower priority.

62. The REC's work was aimed at supporting NGOs. Austria, Canada, Denmark, Finland, Japan, the Netherlands, Norway, and the EC joined the United States as charter countries, along with the recipient countries in the region: Bulgaria, Czechoslovakia, Hungary, Poland, Romania, and Yugoslavia. Eastern European membership now includes all the countries in the region. The donor countries donated around $15 million in start-up funds.

63. The name is the French acronym for "Poland and Hungary Assistance for Restructuring Economies," but the program was quickly extended to include Czechoslovakia and other countries in the region.

64. "Environment Sector Strategy for Central and Eastern Europe," PHARE internal strategy paper, 1991.

65. Durwood Zaelke, CIEL, interview with the author, August 1995.

66. Alex Hittle, Friends of the Earth, interview with the author, November 1992.

67. EBRD, "Environmental Management: The Bank's Policy Approach" (EBRD: London, 1992), 1.

68. Ibid.

69. EBRD, "Revised Environmental Procedures," BDS96-23, March 1996.

70. Ibid.

71. This included the regions of Abruzzi and Molise, Campania, Apulia, Basilicata and Calabria, and Sicily. Indeed, in its first five years, 70 percent of the EIB's 67 projects were in Italy. A few were cofinanced with the World Bank. See various EIB Annual Reports.

72. Treaty of Rome, Article 130, which has become Article 198 E under the Maastricht Treaty. Interestingly, during the EIB's first few years, one member of its board of directors was a World Bank official.

73. Ibid.

74. Specifically, when authorized by the Board of Governors, acting unanimously on a Board proposal. EIB Statute, Article 18.

75. The board of directors' alternates can also take part in board meetings, although they cannot vote unless they are replacing a director. Member states have twelve alternates, and the Commission has its own alternate.

76. The 1992 Maastricht Treaty reaffirmed the Bank's autonomy and important role in promoting the European Union and implementing the Community's policies toward outsider countries. It also enhanced cooperation between the Bank and the Commission on different development funds.

77. Four of these increases occurred as new countries joined the Community.

78. The EIB does have a Directorate for Economics and Information, with a small Economic and Financial Studies office under the Bank's chief economist. However, the research undertaken by this office is related more to the overall strategic positioning of the Bank, than to more active policy work that might impact project design. Alfred Steinherr, chief economist, interview with the author, 27 June 1996.

79. Senior EIB official, interview with the author, 19 June 1996.

80. In interviews with EIB staff, I was repeatedly reminded that the word "development" is not in the Bank's name.

81. The Bank's involvement in ACP countries, which began in 1963, grew out of an initial provision in the Treaty of Rome for member state colonies and territories, which were mainly in Africa. The EIB's activities in the Mediterranean also date back to the early 1960s and stem from cooperation and association agreements aimed at developing closer relations between the EC and those neighboring countries. In addition to offering loans or subsidized loans from its own resources, the EIB also manages risk capital under the Lomé Convention and Mediterranean Financial Protocols. These are funds from member state budgets that the EIB uses to finance conditional loans or equity participation.

82. Commission of the European Communities, *Report on EIB Operations Outside the Community* (Brussels: Commission of the European Communities, 1991), 6.

83. On January 1, 1999, the euro became the legal currency for the eleven participating EU member states, and the EIB and EBRD began using the euro as their reporting currency on that date. Euro rates were based on ECU rates at the time of conversion. The ECU was not legal tender, but rather a unit of account, based on a basket of the EU's currencies. EIB and EBRD figures as of or including 1999 are expressed in euro in this book.

84. Jean-Jacques Schul, interview with the author, 7 March 1995; Ernst-Günther Bröder, interview with the author, 25 March 1997.

85. He was, for example, behind an effort for the Bank to join CIDIE, the Committee of International Development Institutions on the Environment. CIDIE was created in 1980 to supervise MDB implementation of the February 1980 Declaration of Environmental Policies and Procedures Relating to Economic Development. The Declaration was a product of cooperation between the World Bank and UNEP, with the aim of encouraging MDBs to adopt sound environmental lending guidelines. Le Prestre notes that the World Bank wanted to create a level playing field, first, so that its own work on the environment could not be easily undercut by other MDB activities; and second, so that the additional expenditures the Bank made for environmental purposes would not make Bank loans less attractive or competitive vis-à-vis those of other MDBs. CIDIE was never an important actor, however, and it was disbanded in 1995. See Le Prestre, *The World Bank and the Environmental Challenge,* 134–135.

86. Various EIB officials, interviews with the author.

87. EIB, Document 84/4, "Board of Governors: Bank Activity," Report by the Board of Directors, 4 June 1984, 8.

88. Ibid., 9–10.

89. The head of the EIB's Project Directorate admitted the difficulty of making significant changes late in the project cycle, although he emphasized that the technical staff could still refuse to finance the project if they thought environmental issues had not been dealt with properly. See Herbert Christie, "The European Investment Bank: Finance for Environmental Protection in Europe," speech presented at the Earth Summit/United Nations Conference on Environment and Development, Rio de Janeiro, 3–14 June 1992, 3.

90. European environmental policy is based on over 200 directives as well as regulations and six "Environmental Action Programmes." The directives, which allow national authorities to determine the forms and methods of application, have been the most common tool of European environmental policy to date, but are criticized as weak policy instruments, because member nations often side step their obligations by, for example, contesting the scientific basis for decisions. For an analysis, see Juliet Lodge, ed., *The European Community and the Challenge of the Future* (New York: St. Martin's Press, 1993). Regulations are directly applicable to national law and are rarely used for environmental purposes. The action programmes present proposals for legislation that the Commission may pursue and also provide a forum to discuss future directions. Member states have no legal obligation to adopt the action programmes.

91. 1987 was also declared European Year of the Environment by the Council and was the year when the Commission's Fourth Environmental Action Programme was adopted. The SEA contains an entire section on environmental policy, establishing basic objectives, principles, and conditions of EC environmental action without explicitly defining "environment." The objectives include a commitment "to preserve, protect and improve the quality of the environment," "to contribute toward protecting human health," and "to ensure a productive and rational utilization of national resources." The principles focus on preventive action, such as the polluter pays principle. Stanley P. Johnson and Guy Corcelle,

The Environmental Policy of the European Communities, Second ed. (Boston: Kluwer Law International, 1995), 489.

92. EIB, *Environmental Policy Statement,* 4.

Chapter Four: Policy Process: Institutionalizing Environmental Objectives

1. Moises Naim, "The World Bank: Its Role, Governance, and Organizational Culture," C-274.

2. This is a lesson of work by "new institutionalists." See Peter A. Hall and Rosemary C. R. Taylor, "Political Science and the Three Institutionalisms," *Political Studies* 44 (1996): 952–973.

3. Green bankers can be defined institutionally, as in particular jobs that encourage holders to "think green," but it is also true that individual staff committed to green policy can be found in other positions as well.

4. One important exception is Keohane and Levy, *Institutions for Environmental Aid.*

5. Other significant procedures that have an impact on the incentives facing Bank staff include monitoring and evaluation procedures and procurement procedures. I do not examine monitoring and evaluation procedures in depth here, because I am focusing more on the incentives facing staff in project design and development rather than implementation, although one can argue that staff expectations of weak monitoring might move them to include lofty environmental aims no one expects to be achieved. Procurement, in turn, is an under-researched area of study, for which there is little available data. This is an important area for future investigation and analysis.

6. For more on the ways in which a policymaking institution's openness to outside advice affects its ability to implement new ideas and programs, see Margaret Weir and Theda Skocpol, "State Structures and the Possibilities for 'Keynesian' Responses to the Great Depression in Sweden, Britain, and the United States," in Peter Evans, Dietrich Rueschemeyer, and Theda Skocpol, eds., *Bringing the State Back In* (Cambridge: Cambridge University Press, 1985), 107–163.

7. The finance ministers on the boards of governors may also meet outside of these annual meetings. For example, G-7 finance ministers meet more often. EIB governors meet at ECOFIN meetings. A subset of World Bank governors are also members of the IMF–World Bank Development Committee, which meets each spring. The Development Committee, set up in 1974, has twenty-four members, who are usually ministers of finance or development, and they advise the Boards of Governors of both the Bank and Fund on broad policy issues.

8. "Board" refers to the board of directors.

9. This section draws from secondary sources as well as author interviews with the following officials: at the EBRD, executive director officials from the United States, France, Switzerland, Britain, the European Commission, the EIB, Japan,

and Germany, interviewed in April 1995 and June 1996; at the World Bank, ED office representatives of the United States, Britain, Germany, and the multiconstituency office headed by the Netherlands, in May 1996 and April 1997; at the EIB, two members of the management committee, the EIB's former president, and members of the European Commissions DGII, interviewed in June–July 1996 and April 1997. I also interviewed treasury ministry officials in the United States and Britain, in February 1995, July 1995, and July 1996.

10. IDA's membership is slightly smaller, at 159.

11. For example, there is a Swiss ED for the group containing Switzerland, Azerbaijan, Kyrgyz Republic, Poland, Tajikistan, Turkmenistan, and Uzbekistan.

12. The typical EBRD ED office now has an ED, an alternate, and a secretary. A multiconstituency ED office at the World Bank may have a few assistants, but not enough to represent each country in its group.

13. Some of the single constituency offices of major World Bank donors (namely, France and the U.K.) also cover the IMF. They are joint offices that sit on the boards of both institutions.

14. Members of the boards of the World Bank and EBRD often meet informally in smaller groups. For example, at the EBRD assistants to the EU countries' executive directors meet a week before board meetings to discuss projects. There are also occasional board workshops on particular topics, such as potentially contentious loans or the creation of an Energy Efficiency Unit at the EBRD.

15. Treasury has been the dominant agency responsible for instructing United States EDs at the Bank since the late 1960s, when it eclipsed the National Advisory Council on International Monetary and Financial policies (NAC), which then became an advisory group. NAC is chaired by the treasury secretary and also includes leading officials from the Federal Reserve System, the Export-Import Bank, State, and Commerce. For a history and analysis of U.S. participation at the World Bank, see Lars Schoultz, "Politics, Economics, and U.S. Participation in Multilateral Development Banks," *International Organization* 36 (3) (1982): 537–574.

16. U.S. board members admit, off the record, that the latter is clearly a protectionist policy.

17. Richard Gerster, "Accountability of Executive Directors in the Bretton Woods Institutions," *Journal of World Trade* 27 (6) (1993): 115.

18. For analysis on how Congress influences the Treasury's MDB policies, see Jonathan Earl Sanford, "U.S. Policy Toward the Multilateral Development Banks: The Role of Congress," *The George Washington Journal of International Law and Economics* 22 (1) (1988): 1–115. Sanford also shows that Congress has generally been a more active participant with respect to MDB policy than countries with parliamentary or executive-dominated political systems.

19. Formerly the Overseas Development Agency, under the Foreign Office, in May 1997 the DfID was upgraded to become a government department with its own cabinet minister.

20. Gerster, "Accountability of Executive Directors in the Bretton Woods Institutions," 89.

21. Assistant to the U.K. Executive Director, World Bank and IMF, interview with the author, 12 May 1997. U.K. Director's Assistant, EBRD, interview with the author, February 1995.

22. EIB, *Annual Report* (Luxembourg: EIB, 1995).

23. "The World Bank: Its Role, Governance, and Organizational Culture," C-280. Naim was an Executive Director at the Bank from 1990 to 1992.

24. EBRD and EIB annual reports 1994–1996.

25. For analysis of the contributions of World Bank presidents, see Jochen Kraske, William H. Becker, William Diamond et al., *Bankers with a Mission: The Presidents of the World Bank, 1946–91.*

26. Catherine Gwin, "U.S. Relations with the World Bank," 25.

27. Directors from donor countries may also become involved with projects if their countries have bilateral funds that may aid in project development, e.g., in the hiring of consultants to do appraisal work.

28. According to the author's interview with Treasury officials in October 1994, this was due to pressure by major donor countries in the context of the IDA-9 negotiations. It is also worth remembering that while standards of transparency have increased in recent years, for many years the norm for MDBs, like other financial institutions, was that clients' financial data (in this case, a country's macroeconomic data) should be kept secret.

29. Data from 1994–1999 annual reports. EIB data include lending for global loans.

30. For an example of how the Bank promotes judicial reform through its loans and studies, see chapter 3 in Ibrahim F. I. Shihata, *The World Bank in a Changing World: Selected Essays and Lectures,* vol. 2 (Boston: Martinus Nijhoff Publishers, 1995). He discusses loan components that include funding for the drafting of legislation and the publication of legal information.

31. James D. Wolfensohn, "People and Development," Annual Meetings Address 1996, ⟨www.worldbank.org/html/extdr/extme/jdwams96.htm⟩. Although it is too soon to assess the CDF, many view it as positive but also vague since it is ambiguous on implementation issues, it is unclear how it fits into the Bank's operational policies and procedures, it does not prioritize development objectives, and it can easily run into trouble in countries that lack capacity or interest in the encouragement of broader participation. See Deepak Gopinath, "Wolfensohn Agonistes," *Institutional Investor,* September 2000, 43.

32. Naim, "The World Bank: Its Role, Governance, and Organizational Culture," C-273. Wapenhans added the argument that there is also ambiguity in the way new missions are defined, which makes their operational purposes imprecise. "The need for long-term, sensitive, natural resources management has added the concern for sustainable development," argued Wapenhans. However, he adds, "the content of this phrase remains operationally undefined, rendering it vulnerable to the same fate of oblivion that befell such grand schemes as integrated rural

development." The result, he posited, is that interpretation can be so wide as to "defy institutional accountability." Willi A. Wapenhans, "Efficiency and Effectiveness: Is the World Bank Group Well Prepared for the Task Ahead?" in *Bretton Woods: Looking to the Future*, C-292.

33. Shihata, *The World Bank in a Changing World: Selected Essays and Lectures*, 16–17.

34. As a result of Wolfensohn's restructuring, there are now team leaders and program team leaders as well.

35. Gopinath, "Wolfensohn Agonistes," 40. Low morale among staff also prompted Wolfensohn to ask D.C.-based staff why malaise pervaded headquarters. Stephen Fidler, "World Bank Employees Blame Wolfensohn for Morale Crisis," *Financial Times*, 31 January 2001.

36. It also includes staff from the International Finance Corporation (IFC).

37. Wapenhans, "Efficiency and Effectiveness: Is the World Bank Group Well Prepared for the Task Ahead?" C-289–304. Willi A. Wapenhans, *Effective Implementation: Key to Development Impact*, Portfolio Management Task Force. Confidential Report of the World Bank (Washington, D.C., World Bank, 22 September 1991), iii.

38. Rich, *Mortgaging the Earth*, 183.

39. Wade, "Greening the Bank," 717.

40. It is also worth noting that the Bank's analytical work on environmental issues has declined in terms of being dated. According to the Bank, "during 1995–99, analytical work less than five years old was available for only 14% of countries." World Bank, *Making Sustainable Commitments: An Environment Strategy for the World Bank* (Washington, D.C.: World Bank, 2001), 25.

41. Quote from author interview with G-7 ED official at World Bank, 25 March 1997. Among the government officials praising Bank research are environment ministry officials in Hungary and Poland, interviewed by the author, October 1997.

42. World Bank environmental Task Manager, ECA department, interview with the author, 15 July 1997.

43. World Bank, *Making Sustainable Commitments*, 43.

44. The World Bank Group's IFC is a private sector lending institution that has a different set of clientele and is largely demand driven, more like the EBRD. However, since it is somewhat separate from the IBRD, and since it is a relatively small player in Central and Eastern Europe, I do not count it among my MDB cases.

45. The amounts are not insignificant. In fiscal 1999, for example, trust fund disbursements totaled $1.3 billion. Trust funds consist of contributions from developed country donors and vary in terms of the extent to which they are a form of tied-aid. In fiscal 1999, the largest contributors were the Netherlands ($221 million), Japan ($199 million), the United States ($94 million), the United Kingdom ($87 million), and Sweden ($74 million). World Bank, *Annual Report*

(Washington, D.C.: World Bank, 1999), 144. The EBRD also has "technical co-operation funds" used for project preparation and implementation, but the amounts are smaller. Commitments totaled around EUR 90 million for 1999, for example. EBRD, *Annual Report* (London: EBRD, 1999), 56.

46. World Bank, "The Strategic Compact: A Summary Note," 12 July 1997, ⟨http://www.worldbank.org⟩.

47. ODs, in turn, evolved from operational manual statements (OMSs) and operational policy notes (OPNs), both written instructions from management for staff to follow. These were supplemented a few years later by operational policy statements, bank procedures, and good practice statements. Shihata, *The World Bank in a Changing World*, 546–547.

48. One example is the Bank's Resettlement Policy, which was issued in 1980 and revised a few times. NGOs are highly critical of it. See Rich, *Mortgaging the Earth*, 156.

49. For more details on each stage of the project cycle, see Warren C. Baum, *The Project Cycle* (Washington, D.C.: The World Bank, 1982).

50. Author interviews with Bank staff, 25 July 1995.

51. While NEAPs have been required for IDA borrowers since the late 1980s, many countries have yet to undertake them. The Bank encourages IBRD members also to complete NEAPs, but this is not a requirement. "Effectiveness of Environmental Assessments and National Environmental Action Plans: A Process Study," Operations Evaluation Department Report 15835, 28 June 1996, 3.

52. Office of the Senior Vice President, Development Economics, "Integrating Environmental Issues into Country Assistance Strategy," 1 February 1996, v, 14, cited in Wade, "Greening the Bank," 721.

53. World Bank, "Promoting Environmental Sustainability in Development: An Evaluation of the World Bank's Performance," Operations Evaluation Department, Sector and Thematic Evaluation Group, 15 September 2000, 18.

54. World Bank, "Promoting Environmental Sustainability in Development: An Evaluation of the World Bank's Performance," 21.

55. Wade, "Greening the Bank," 722.

56. World Bank, "Toward an Environment Strategy for the World Bank Group." Progress Report/Discussion Draft, April 2000, 25.

57. Katrina Brandon, "Environment and Development at the Bretton Woods Institutions," in *Bretton Woods: Looking to the Future*, C-137.

58. For case studies on how well the Bank has complied with its policies on indigenous peoples, involuntary resettlement, water resources, information disclosure, and the Inspection Panel, see Jonathan A. Fox and L. David Brown, eds., *The Struggle for Accountability: The World Bank, NGOs, and Grassroots Movements* (Cambridge: The MIT Press, 1998).

59. "BP" stands for "Bank procedures."

60. This was in fiscal 1999, reflecting in part loans to help countries in the midst of the Asian financial crisis.

61. Operations Evaluation Department, "Effectiveness of Environmental Assessments and National Environmental Action Plans," 24. This study is based on research conducted in eight countries, of which one (Poland) was in the CEE region.

62. Ibid., 6.

63. Ibid., 26.

64. Ibid., 6.

65. Ibid., 45–62.

66. World Bank, "Making Sustainable Commitments: An Environment Strategy for the World Bank," 6.

67. World Bank, *Annual Report* (Washington, D.C.: World Bank, 1997), 14–15.

68. At the same time, it's worth noting that the Bank under Wolfensohn also has a low threshold of tolerance for criticism from within its ranks, as witnessed by the resignation of Joseph Stiglitz, its outspoken chief economist, in 1999. Stiglitz resigned amid objections to his public criticism of the way the IMF and the U.S. Treasury handled the Asian and Russian financial crises in the late 1990s. A year later, Ravi Kanbur resigned as the staff director of the Bank's 2000 World Development Report, *Attacking Poverty,* after he felt pressured to change the report's main messages to be more in line with free-market thinking. See Robert Wade, "Showdown at the World Bank," *New Left Review* 7 (January/February 2001).

69. Lori Udall, "The World Bank and Public Accountability: Has Anything Changed?" in *The Struggle for Accountability,* 406–407.

70. Member of EMTEN staff, World Bank, interview with the author, 9 January 1996.

71. Nelson, *The World Bank and Non-Governmental Organizations,* 2.

72. Jane G. Covey, "Is Critical Cooperation Possible? Influencing the World Bank Through Operational Collaboration and Policy Dialogue," in *The Struggle for Accountability,* 87.

73. David Hunter, "The Role of the World Bank in Strengthening Governance, Civil Society, and Human Rights," in *Lending Credibility: New Mandates and Partnerships for the World Bank,* ed. Peter Brossard et al. (Washington, D.C.: World Wildlife Fund, 1996), 67.

74. Draft letter from Bruce Rich and Stephanie Fried, Environmental Defense Fund, to James Wolfensohn, August 6, 1998. They cite Marcus W. Brauchli, "World Bank Is Hurt by its Failure to Anticipate the Indonesia Crisis," *Wall Street Journal,* 14 July 1998, 1, which notes that the Bank knew that corruption in its Indonesian projects was widespread, but it avoided directly addressing the issue partly because it did not want to confront the government.

75. Ibid., 74.

76. See memo by Kay Treakle, Bank Information Center, 6 April 1999, in Bank Information Center, "Tuesday Group Report," April 1999, volume 3 (4). For

case studies on the Inspection Panel, see Udall, "The World Bank and Public Accountability: Has Anything Changed?" 416–421, which discusses the Arun III claim; and Margaret E. Keck, "Planafloro in Rondônia: The Limits of Leverage," in *The Struggle for Accountability,* 181–218, which discusses the Planafloro project designed to address some of the environmental problems caused by the Bank-financed Polonoroeste project.

77. EBRD, "Environmental Management: The Bank's Policy Approach" (London: EBRD, 1992).

78. Interviews with EBRD staff, April 1995, August 1996.

79. For example, the municipal and environment team has worked closely with the Project Preparation Committee (PPC), a network of bilateral donors and the MDBs that act as matchmakers to bring together bilateral and multilateral assistance for projects that support the Regional Environmental Action Programme. The EBRD contains the PPC's small secretariat. The PPC's handful of officers are split between the World Bank and EBRD. The PPC is described in greater depth in chapter 5.

80. Senior EBRD banker, interview with the author, 9 July 1996.

81. Author interviews with EBRD board members, June 1996.

82. EBRD, *Annual Report* (London: EBRD, 1996), 30.

83. Member of the EBRD's municipal team, interview with the author, 15 February 1995.

84. "EBRD and Energy Efficiency" (London: EBRD, 1998). Again, shareholder pressure was key in the unit's creation; in this case, the major mover behind its birth was the United States Executive Director at the time, Jim Scheuer, a retired United States congressman (Democrat) who was keen on the promotion of energy efficiency. Scheuer was also involved in efforts to reform the World Bank's environmental behavior in the 1980s. See Rich, *Mortgaging the Earth,* 121–122.

85. EBRD, "Environments in Transition: The Environmental Bulletin of the EBRD" (London: EBRD, 1997), 2–6.

86. Donor pledges to the five funds exceeded 1.5 billion euro by mid-2000. EBRD, *Annual Report* (London: EBRD, 2000), 62.

87. Ibid., 22.

88. Two examples of this are the Maritza East Power Project and the Transit Roads Project in Bulgaria. According to Goldberg, Bulgarian national law would also have required these projects to receive a full EIA. The Bank's own screening categories call for thermal power plants with a heat output of 300 megawatts or more, as well as the construction of highways and roads to receive a full EIA. Donald M. Goldberg and et al., "The European Bank for Reconstruction and Development: An Environmental Progress Report" (Washington, D.C.: CIEL, 1995), 24.

89. Ibid., xxvii–xviii.

90. Interviews with EBRD representatives in Hungary, finance and environmental ministry officials in Hungary and Poland, and a central bank official in Hungary, October–November 1997.

91. EBRD, *Annual Report* (London: EBRD, 1999), 20.

92. The 5 million euro rule appears to be flexible, since there are a number of projects under this amount, particular in areas involving an EBRD equity stake.

93. I am referring to a 1994 loan to Danone-Serdika SA in Bulgaria to upgrade and modernize its product line; 1993 loans to Coca-Cola Tirana in Albania and Coca-Cola Bihor & Iasi in Romania to build soft drink bottling facilities; and at least six loans for hotel development in the CEE region.

94. This was Stephen Lintner, whose role as a World Bank "green banker" is discussed in chapter 5.

95. Interviews with officials in U.S. Treasury MDB office, Washington, D.C., 28 March 1996; German ED official, London, 16 June 1996.

96. Official in United States ED office, London, interview with the author, 15 January 1997.

97. Some board members, including the Swiss and French, were also not in favor of greater information disclosure. Author interviews with various ED offices, June 1996.

98. Official from U.S. Treasury MDB office, Washington, D.C., interview with the author, 28 March 1996.

99. "EBRD Increases Public Access to Information," Press Release, 1 July 1996.

100. Some examples include the addition of a nature preserve as part of Bulgarian highway loan; the addition of a PHARE-funded bicycle path as part of a project in Kaunas, Lithuania; and so on. Interviews with EBRD staff, June 1996, and NGOs in the Baltic States, May 1998.

101. Goldberg, "The European Bank for Reconstruction and Development: An Environmental Progress Report," ix.

102. Ibid., 11.

103. Ibid., 12.

104. Ibid., xii.

105. CEE Bankwatch Network, "K2/R4 Completion Projects," ⟨http://www.ecn.cz/k2r4/BOARD.sTM#2.2⟩.

106. Personal e-mail communication from Bulgarian NGO, Centre for Environmental Information and Education, 16 February 1998. Personal e-mail communication from Jozsef Feiler, Policy Coordinator, CEE Bankwatch Network, 22 July 1998.

107. For example, the World Bank generally avoids lending for expensive end-of-the-pipe technology in Central and Eastern Europe, such as flue-gas-desulphurization, preferring to look at other policy and project alternatives. The EIB, on the other hand, will lend money for such technology.

108. There are also other signs that the Bank's mission is expanding. As a result of the Amsterdam Summit in 1997, the EIB drew up a program to invest in new areas of health and education, as well as to do more in the areas of urban renewal and environmental protection.

109. Department for International Development, "Working in Partnership with the European Investment Bank," March 2000, 2.

110. An exception is the Bank's new preaccession facility, which is not backed by the EU guarantee.

111. Annex 1 projects include crude oil refineries; power stations and combustion plants generating more than 300MW; installations for the storage and disposal of radioactive waste; foundries; installations for asbestos extraction; integrated chemical installations; motorways, express roads, long distance railway lines, and airports of a certain size; port vessels larger than 1,350 tons; and installations to dispose of waste produced by incineration, chemical treatment, or land fill of toxic, dangerous wastes.

112. Annex 2 projects include a broader range of investments in sectors including agriculture, energy, rubber, metals, chemicals, food, and infrastructure. Examples in the energy sector include surface storage of natural gas, power installations not included in Annex 1, underground combustible gas storage, and installations that produce and enrich nuclear fuels or reprocess irradiated nuclear fuels.

113. Senior EIB officials, interviews with the author, June 1996.

114. See, for example, Tamara Raye Crockett and Cynthia B. Schultz, "The Integration of Environmental Policy and the European Community: Recent Problems of Implementation and Enforcement," *Columbia Journal of Transnational Law* 19 (169) (1991):181–182; Alberta Sbragia, "EC Environmental Policy: Atypical Ambitions and Typical Problems?" in *The State of the European Community: The Maastricht Debates and Beyond,* ed. Alan W. Cafruny and Glenda G. Rosenthal (Boulder: Lynne Rienner Publishers, 1993), 337–352. To illustrate the problem of weak enforcement, in 1996, for example, the Commission registered over 600 environmental complaints and infringement cases against member states. See "Implementing Community Environmental Law," Communication to the Council of the European Union and the European Parliament (Brussels: European Commission, 1996). Beginning in 1997, the European Commission began fining member states for noncompliance with European Court of Justice judgments on environmental law infringements.

115. One goal of the 1997 amendment of the EIA directive was to reduce member state wiggle-room on addressing Annex II projects by clarifying the circumstances under which these projects are required to have an EIA undertaken.

116. Barbara Connolly and Tamar Gutner, "Organizational Inertia and Innovation: Environmental Aid to Central and Eastern Europe."

117. As EIB official Peter Carter pointed out in an April 1998 correspondence to the author, it can either walk away from the project or request certain corrective actions.

118. European Commission, "Working Procedures between the EIB and the Commission Services (DGXI and XXII) in the Consultation of the Commission under Article 21 of the EIB Statute," 30 October 1992.

119. Author interviews with EIB officials, and DGXI officials, June–July 1996. For the general procedure for the Commission's opinions on EIB operations, see

the most recent version of the "Vademecum: Procedures for the Commission's Opinion on EIB Operations," European Commission, DGII.

120. Author interviews with EIB officials, June 1996.

121. James N. Barnes and Sandrine Bretonni, "An Overview of the European Investment Bank: Accountability and Transparency" (Paris: Friends of the Earth, 1997).

122. Richard Douthwaite, "Why It Matters Whom Ireland Sends to the European Investment Bank," Feasta, the Foundation for the Economics of Sustainability, ⟨http://sustainable.buz.org/feasta_pages/article-EIB.html⟩.

123. Peter Conradi, "EU Bank 'Lost £180m on the Bond Market,'" *Sunday Times,* 12 September 1999, accessed at http://www.Sunday-ti...ages/sti/99/09/12/stifgneur02003.html?999. This followed charges by a former EIB employee that the Bank lost these millions by speculating, and that they also used brokers that had personal connections to it, even though they charged significantly higher commissions than their competitors. The EIB countered that these criticisms were the result of a legal dispute between the ex-employee and the Bank.

124. European Investment Bank, "Rules on Public Access to Documents," Adopted by the Bank's Management Committee on 26 March 1997, 97/C 243/06.

Chapter Five: MDB Environmental Policies and Practice in CEE

1. It is worth pointing out that commercial banks have not played a significant role in providing environmental financing to CEE or the former Soviet states. This is attributed to a low demand for such financing, few attractive lending opportunities available to commercial banks, and the absence of knowledge within the banks on how to design such financing. OECD, "Report on Environmental Financing in CEEC/NIS," Task Force for the Implementation of the Environmental Action Programme in Central and Eastern Europe, document CCNM/ENV/EAP (98)24/REV2, 18 May 1998, 11.

2. The term "donors" here refers to bilateral *and* multilateral donors. For some aid officials, "donors" is synonymous with bilateral agencies and does not include the MDBs.

3. This figure includes loans, World Bank IDA no-interest loans, and EBRD equity financing and guarantees. As noted in chapter 1, this figure includes Central European countries and the three Baltic states. See table 5.1 for details. For comparison, the largest single donor to the region is Germany, which committed around $20 billion to CEE and NIS countries in the period 1990–1999. Following Germany is the EU's PHARE program, which committed around 8.9 billion euro to CEE in the period 1990–1998, and the U.S. SEED (Support for East European Democracy) program, which appropriated a total of $5.2 billion to CEE alone in 1990–1999. The bulk of German, U.S., and PHARE money has been in the form of grants. Sources: Phare 1998 Annual Report; "SEED Act Implementation Report" Fiscal Year 1997, U.S. Department of State, February 1998, table 1;

USAID Congressional Presentations, FY 1990–2001; "Aid at a Glance," www.oecd.org/dac/htm/agdeu.htm.

4. Barbara Jancar, "Democracy and the Environment in Eastern Europe and the Soviet Union," *Harvard International Review* 12 (4) (1990): 14.

5. These were primarily in the southern and western parts of the country. Tomasz Zylicz, "In Poland, It's Time for Economics," *Environmental Impact Assessment Review* 14 (2,3) (1994): 79–94. These areas were actually first designated as ecological hazards by the Polish Council of Ministers in 1983. Clyde Hertzman, "Environment and Health in Central and Eastern Europe: A Report for the Environmental Action Programme for Central and Eastern Europe" (Washington, D.C.: The World Bank, 1995), 6.

6. World Bank, EMTEN, "Environmental Policy in Central and Eastern Europe," 19 November 1991.

7. World Bank, "Environmental Action Programme for Central and Eastern Europe," document submitted to the Ministerial Conference in Lucerne Switzerland, 28–30 April 1993 (Washington: World Bank, 30 June 1994), 9 (hereafter cited as "EAP"). Among the health problems associated with environmental degradation in northern Bohemia are reduced life expectancy compared with the rest of the country and comparatively high mortality rates for lung cancer, infant mortality, respiratory and cardiovascular diseases. Hertzman, "Environment and Health in Central and Eastern Europe," 19.

8. Lead poisoning is associated with learning, behavioral, and hearing problems in children.

9. EAP, 9. Descriptions of the health problems associated with these areas are in Hertzman, "Environment and Health in Central and Eastern Europe," 19–20.

10. "Cleaning Up After Communism," *Economist,* 17 February 1990, 54.

11. Penn Kemple, "East Europe: Greening of the Reds," *Washington Post,* 24 December 1989, C3.

12. Hertzman, "Environment and Health in Central and Eastern Europe," x.

13. Schreiber, "The Threat From Environmental Destruction in Eastern Europe," 361.

14. World Bank, "Environmental Policy in Central and Eastern Europe," 20.

15. Cited in European Commission, "Environment for Europeans," no. 1, March 2000, 5.

16. Definitions for levels of treatment are not precise. Primary treatment usually refers mainly to physical processes, such as gravity settling, to take out large particles. Secondary treatment generally adds biological processes to reduce organic materials. Tertiary treatment adds another level of chemical, physical, or biological processes to further reduce inorganic and organic materials. On CEE data, see Axel Hörhager, "An Environmental Perspective for Eastern Europe," *EIB Papers* 18 November (1992): 39.

17. Jürg Klarer, Bedrich Moldan, eds., *The Environmental Challenge for Central European Countries in Transition* (New York: John Wiley and Sons, 1997).

18. Zbigniew Bochniarz, Representative of Polish Ecological Club and Senior Fellow, Hubert H. Humphrey Institute of Public Affairs, University of Minnesota, hearing before Subcommittee on Transport and Hazardous Materials of the Committee on Energy and Commerce (U.S. Environmental Initiatives in Eastern Europe), U.S. House of Representatives 101st Congress, 23 April 1990, Serial No. 101-140.

19. Stanley J. Kabala, "Environment and Development in the New East Central Europe: Addressing the Environmental Legacy of Central Planning" (Middlebury, VT: Geonomics Institute, 1991).

20. Richard N. L. Andrews, "Environmental Policy in the Czech and Slovak Republic," in *Environment and Democratic Transition: Policy and Politics in Central and Eastern Europe,* eds. Anna Yari and Pal Tamas (Boston: Kluwer Academic Publishers, 1993), 5–48.

21. This was a joint project between Czechoslovakia and Hungary that entailed the construction of three new dams on the Danube (two on the Czechoslovak side and one on the Hungarian side) and two hydroelectric power plants. The project also entailed creating a new reservoir and rerouting the Danube for 25 kilometers via a new canal that would sharply slow the flow of the river. Hungarian environmentalists were concerned that the project would have negative impacts on agriculture and drinking water, among other things, and with growing vocal protest, the movement resulted in the Hungarian government agreeing to halt work on the dam temporarily in 1981, and finally permanently, in 1989. However, Slovakia continued constructing a dam on its side of the border. See Judit Galambos, "Political Aspects of an Environmental Conflict: The Case of the Gabcikovo-Nagymaros Dam System," in *Perspectives on Environmental Conflict and International Relations,* ed. Jyrki Kakonen (London: Printer Publishers, 1992), 72–95.

22. *Economist,* "Clean Up or Clear Out," 11 December 1999, 47. The ten accession countries are: Bulgaria, Czech Republic, Estonia, Hungary, Latvia, Lithuania, Poland, Romania, Slovak Republic, and Slovenia.

23. Ibid.

24. *Green Horizon,* vol. 2 (8), October 1999.

25. Gordon Hughes, "Is the Environment Getting Cleaner in Central and Eastern Europe? Selected Evidence for Air Pollution and Drinking Water Contamination" (Washington, D.C.: The World Bank, 1995), 5.

26. OECD, "Environmental Indicators: A Review of Selected Central and Eastern European Countries" (Paris: OECD, 1996).

27. Hughes, "Is the Environment Getting Cleaner in Central and Eastern Europe?" 11. He uses as an example discharges of arsenic in Bulgaria's Maritza River basin. Discharges of arsenic from copper smelters have declined, but this is not reflected in the levels of arsenic in parts of the river near these smelters.

28. Schreiber, "The Threat From Environmental Destruction in Eastern Europe."

29. Julia Bucknall, "Poland: Reducing the Costs of Complying with EU Environmental Legislation," in *Rural Development, Natural Resources and the Environment,* ed. L. Alexander Norsworthy (Washington, D.C.: World Bank, 2000), 85, 89.

30. It appears it left it up to donors to classify their commitments. OECD, "Environmental Financing in CEEC/NIS," June 1998, http://www.mem.dk/aarhusconference/issues/Finance/ececep50.htm.

31. Ibid.

32. Commitments also do not include cancellations and other adjustments.

33. Ibid. For a comprehensive account and analysis of the Environment for Europe process, see Connolly and Gutner, "Organizational Inertia and Innovation."

34. Ibid.

35. Ibid., 1. The Environment for Europe process grew out of a campaign in the early 1990s by Josef Vavrousek, then the Czechoslovak federal environment minister, to set up a permanent Council of European Environmental Ministers who would hold regular conferences, set action plans and targets, and formulate strategies and targets to implement them. Vavrousek found that his counterparts in Western Europe were not eager to create new institutions, but they were willing to meet occasionally to discuss East-West environmental issues, with an emphasis on managing environmental aid and developing environmental policy.

36. "Policy network" has been defined by a number of political scientists as clusters of public and private actors who organize around particular topics to pursue their policy goals. See Tamar Gutner and Stacy D. VanDeveer, "The Role and Impacts of Mature Policy Networks in the Baltic Sea Region" (paper presented at the annual meeting of the International Studies Association, Chicago, Ill., February 2001).

37. Following the first "Environment for Europe" meeting in 1991, European environmental ministers asked the World Bank and OECD to draft a broad framework that identifies priorities for environmental reform and expenditures in CEE. The OECD assisted the Bank in producing the document, which included input from a variety of donor and recipient governments, as well as a number of supporting studies undertaken by consultants.

38. The Task Force, based at the OECD, is a small secretariat whose goal is to help CEE states implement policy and institutional aspects of the EAP. Its members include representatives from both donor and recipient countries, as well as international organizations and several NGOs. The Task Force's job is to help recipient countries adapt EAP recommendations to local conditions through the development of national environmental action programs (NEAPs). It also coordinates efforts with the PPC in assisting countries to better mobilize financial resources to address important environmental problems. The Task Force also hosts meetings for recipients, workshops on specific topics of environmental management, and other activities that bring donors and recipients together.

39. European Commission, "The Weight of the Law," *Environment for Europeans,* March 2000, 13.

40. The World Bank and EBRD provide other useful business data statistics, while the EIB does not. For example, for the World Bank, net disbursements for the countries in table 5.1, between 1991–1999 totaled $6 billion, compared with $14.1 billion in commitments. The EBRD has data for net cumulative business volume, which equals commitments minus cancellations. For the same period, its net cumulative business volume totaled 7.5 billion euro, compared with 8.3 billion euro in commitments. World Bank, "Net Disbursements, IBRD and IDA Loans." At http://www.lnweb18.worldbank.org/eca/eca.nsf...dff47e01a338525 69d00076392a>OpenDocument. EBRD data is from author correspondence with Josue Tanaka, 1 February 2002.

41. The relationship between the Bank and many of the countries in the region predates the regime changes of the early 1990s. Poland, Czechoslovakia, and Yugoslavia were among the founding members of the Bank, although Poland and Czechoslovakia pulled out a few years later. Romania has been a member of the World Bank and IMF since 1972, although its communist leadership stopped borrowing in 1982. Hungary joined the Bank in 1982, and Poland rejoined the Bank in 1986. Other countries in the region are more recent members.

42. Various annual reports; "The World Bank Streamlines Its Strategy for Transition Economies," *Transition* (Washington, D.C.: World Bank, February 1997).

43. Loans that fall under "miscellaneous" include those that provide foreign exchange for imports, landmine clearance, and soldier demobilization in Bosnia and Herzegovina.

44. EAP, II-26.

45. GEF projects are not analyzed here because the GEF is a separate facility involving the World Bank, UNDP, and UNEP that operates with grant funding. Although worth mentioning as an important component of the World Bank's environmental activity, it is not directly comparable with work undertaken by the other two MDBs.

46. It is worth noting that this list is more generous than one of the Bank's own figures, which puts its cumulative lending for environmental purposes in the ECA region at $470.5 million. World Bank, *Annual Report* (Washington, D.C., World Bank, 1999), 217.

47. The project predicted it would reduce air pollution in the country on an annual basis from 1995 to 2010 by 240,000 tons of ash, 50,000 tons of SO_2, 8,000 tons of carbon monoxide, 20,000 tons of nitrogen oxide, and 3.1 million tons of carbon dioxide. World Bank, "Staff Appraisal Report: Poland Energy Resource Development Project" (Washington, D.C.: World Bank, 1990).

48. The staff appraisal report estimated that the project would save $20–$40 million a year in energy costs, while reducing SO_2 emissions in northern Bohemia by 20 percent, or about 10 percent for the Czech Republic. World Bank, "Staff Appraisal Report: Czech and Slovak Republic Power and Environmental Improvement Project" (Washington, D.C.: World Bank, 1992).

49. C'EZ is the largest company in the Czech Republic with 12,000 employees, and it generates around 80 percent of the country's electricity. European Invest-

ment Bank, "C'EZ Power Plant Improvement" (Luxembourg: European Investment Bank, 1995).

50. The wet limestone FGD equipment was installed at the Prunerov II power plant, which was the single largest source of SO_2 in the country. Ibid., 32, 35.

51. District heating systems consist of boilerhouses that heat water, which is piped out to recipients, such as households.

52. World Bank, "Staff Appraisal Report: Estonia District Heating Rehabilitation Project" (Washington, D.C.: World Bank, 1994).

53. The GIS is the source of databases with geographic, geologic, economic, and other information to help in land-use planning and environmental protection issues.

54. Helmut Schreiber was the environmental coordinator for EC2, the country department working on Albania, Bosnia and Herzegovina, Croatia, Czech Republic, Hungary, Poland, Slovakia, and Slovenia. After the mid-1997 reorganization, he was a team leader for Slovenia, Hungary, the Czech Republic, and Slovakia for environmental projects.

55. Official of UAB Geoterma, Klaipeda, Lithuania, interview with the author, May 25, 1998. The heat from the geothermal water is used to preheat the district heating water, which reduces the heating plant's need to use oil.

56. The three Baltic states are signatories of the 1974 "Convention on the Protection of the Marine Environment of the Baltic Sea Area" (the Helsinki convention), which requires them to address marine pollution and to implement a number of other recommendations developed by the HELCOM, the Helsinki Commission. The JCP grew out of this convention. The World Bank played an important role in the development of the JCP, which was signed by environmental ministers of the Baltic Sea region.

57. Stephen Lintner was also a key actor in devising the EIA policies of the World Bank and the EBRD. In addition, he also helped to draft the JCP for the Baltic Sea.

58. A key official for this project was Richard Ackermann, who was also a leading player in drafting the EAP.

59. In 1993, the Czech Prime Minister, Vaclav Klaus, was particularly critical of the World Bank and EBRD, saying they were "pushing us to accept loans, standard loans, which is absolutely wrong. And we had tremendous troubles to reject all of them." He argued that the country needed first to focus on price liberalization, trade liberalization, and currency convertibility. Vaclav Klaus, "Remarks of the Czech Prime Minister Vaclav Klaus to the Bretton Woods Committee," Washington, D.C., 15 October 1993.

60. Senior World Bank environmental economist, interview with the author, February 1995.

61. This section is based on personal interviews with municipal water company officials, Bank project implementation units, and municipal officials in Haapsalu, Estonia; Liepaja, Latvia; and Siauliai and Klaipeda, Lithuania; as well as environmentalists and environmental ministry officials in all three Baltic states, June 1998.

62. Official of Haapsalu Water Works, Haapsalu, Estonia, interview with the author, 15 May 1998. "Staff Appraisal Report: Republic of Estonia: Haapsalu and Matsalu Bays Environment Project," World Bank, 15 March 1995.

63. The city's sludge is pumped directly into lagoons, which were largely filled by 1996, and sludge was not distributed according to levels of toxicity. World Bank, "Staff Appraisal Report: Republic of Lithuania Siauliai Environment Project," World Bank, 1995, 10.

64. The old sewage treatment plant could process only 60 percent of daily collected water, which meant a daily discharge of around 10 tons of polluted water into the Kulpe River, which flows directly into the Baltic Sea. The sewage is highly contaminated with heavy metals, which also means that the sludge removed from the plant cannot be used for agriculture. "Ecological Problems of Siauliai," Siauliai Municipality Environmental Affairs Department, 1994, 12, 15.

65. In fact, Siauliai is the source of most of the pollutant load discharged into the Lielupe. Pollution sources on Latvia's portion of the river tend to be small.

66. "Ecological Problems of Siauliai," 12, 15.

67. According to the project's task manager, two of the big pig farms, with 20,000 pigs each, contribute more waste than the entire city of Siauliai. World Bank project manager, interview with the author, 30 April 1998.

68. World Bank Quality Assurance Group, Draft synthesis document, 24 April 1997, 5 (hereafter cited as "QAG Report").

69. Albert O. Hirschmann, *Development Projects Observed* (Washington, D.C.: The Brookings Institution, 1967), 1–3.

70. In July 1998, for example, the World Bank hired outside auditors to investigate what it called "alarming information" about possible fraud in the form of embezzlement and kickbacks. Lorraine Adams, "World Bank Hires Auditors to Probe its Own Spending," *Washington Post,* 16 July 1998, A1, 21.

71. Of course, even in this area, the Bank is not without its critics. See, for example, the Meltzer Commission Report, which gives the World Bank's evaluation process "low marks for credibility," arguing that the measure of sustainability receives too small a weight in evaluation while results are measured only after funds are disbursed. International Financial Institution Advisory Commission (Meltzer Commission) report. (Washington, D.C.: International Financial Institution Advisory Commission, March 2000). Available at http://phantom-x.gsia.cmu.edu/IFIAC/USMRPTDV.html, 6.

72. See World Bank information on Poland, ⟨http:wbln0018.worldbank.org/ECA/eca.nsf/27cf96b3049c241c852567d100133985/255f2d599cec8b24852567ef00590287?OpenDocument⟩.

73. Projects in extended problem status included: in Bulgaria, a 1993 technical assistance project, a 1994 agricultural development project, and the 1994 water company restructuring project (which, as noted above, had positive environmental impacts); in Hungary, a 1992 project for highway rehabilitation, a 1993 proj-

ect in health service intervention and training, and a 1993 project to reform social insurance; and in Poland, a 1992 housing and urban development project. Prem Garg, Office Memorandum, 13 May 1997, "Portfolio Monitoring Report."

74. World Bank, "Memorandum of the President of IBRD and IDA to the Executive Directors on a Country Assistance Strategy of the World Bank Group for Hungary," 1998.

75. Ibid.

76. QAG Report, section on oil and gas, i.

77. The physical objectives referred to project-supported equipment, almost all of which was installed and operating by the time the report was written. These objectives were exceeded owing to cost savings and additional funding from other domestic actors and bilateral donors. World Bank, "Implementation Completion Report: Poland Environment Management Project" (Washington, D.C.: World Bank, 30 May 1997).

78. Luis Landau and Julius Gwyer, "Poland Country Assistance Review: Partnership in a Transition Economy" (Washington, D.C.: World Bank, 1997), 11. This is the only such comprehensive report undertaken to date in the region.

79. Ibid., 2.

80. Ibid., 4.

81. Ibid., 95.

82. Ibid.

83. Ibid., 98.

84. In the fall of 1997 and spring of 1998, I interviewed: in Poland, officials from the finance ministry, environmental ministry, National Fund, World Bank and EBRD resident missions, Institute of Sustainable Development, and CEE Bankwatch; in Hungary, officials from the central bank, environmental ministry, finance ministry, World Bank and EBRD environmental missions, a former environmental policy official from Slovenia who is director of the Budapest-based Regional Environmental Center (REC), other officials from REC, World Bank and EBRD environmental officials, environmentalists from the Clean Air Action Group, Energy Club, and CEE Bankwatch; in the three Baltic states, I also interviewed officials from finance and environmental ministries, resident MDB mission officials, CEE Bankwatch and local NGOs, as well as municipal officials, project managers, and other project officials for the three MDBs' water projects and many of their energy projects.

85. Official of World Bank's Warsaw office, interview with the author, 15 October 1997.

86. Official of UAB Geoterma, Klaipeda, Lithuania, interview with the author, 25 May 1998.

87. Linas Vainius, Lithuania Green Movement, Vilnius, Lithuania, interview with the author, 20 May 1998.

88. CEE Bankwatch Network, "Mail," March 2000, 2.

89. PHARE eventually agreed to finance the transformer. Author interviews with World Bank official, Vilnius, Lithuania, 29 May 1998; and World Bank official, Washington, D.C., 30 April 1998.

90. World Bank official, Vilnius, Lithuania, interview with the author, 29 May 1998.

91. This figure excludes credit lines and other multiproject facilities that focus solely on the non-Baltic former Soviet states.

92. For example, under the 70 million ECU facility the EBRD financed with the Swiss company, Landis & Gyr (Europe) Corp., the EBRD provides 35 percent of the financing for each ESCO set up by Landis & Gyr, which in turn provides the balance and takes a majority stake in the ESCO.

93. EBRD, "East European Energy Wastage to be Tackled by EBRD and Landis & Gyr Partnership," Press Release (London: EBRD, 18 December 1996).

94. This would include the reconstruction of boilers to burn more energy efficient fuels, to use geothermal energy for heating purposes, to manufacture briquettes from waste wood, and so on. EBRD, "Former Yugoslav Republic of Macedonia: Power Sub-Sector Project," Report to Board of Directors, BDS93-157, 30 November 1993.

95. This included financing for insulation for 300,000 apartments, hospitals, and schools; refurbishing district heating substations in five cities and towns, by installing heat controls and meters; and purchasing spare parts, insulation material, and various instruments to improve energy efficiency. EBRD, "Estonia Energy Sector Emergency Investment Project," Report to Board of Directors, EDS92-124, 9 November 1992.

96. Member of EBRD energy efficiency team, interview with the author, 12 June 1996.

97. The Municipal and Environmental Infrastructure Unit at the Bank is also one of the top users of the Bank's technical cooperation funds, bilateral pots of grant money provided to the Bank to fund technical cooperation activities, including project preparation, project implementation, advisory services, sector studies, and training. Head of EBRD Municipal and Environmental Infrastructure Unit, interview with the author, July 1996.

98. CEE Bankwatch was founded in 1995 in Poland. Its members include environmentalists throughout the region. Its goal is specifically to monitor the MDBs while raising public awareness of MDB activities and promoting public participation in MDB decision making. "The CEE Bankwatch Network for Monitoring International Financial Institutions," Krakow, 4 March 1997.

99. "E-note," International Institute for Energy Conservation, August 1997.

100. "Is Nuclear Safety Account Going to Fail?" Press Release, CEE Bankwatch, March 1998.

101. EBRD, "Independent Safety Panel's Recommendations on Ignalina NPP Released by Nuclear Safety Account," 21 March 1997, http://www.ebrd.com/english/opera/nucsafe/prelease/18mar21.htm.

102. EBRD, "Estonia: Tallinn Water and Environment Project," Report to Board of Directors, BDS94-93, 24 June 1994.

103. Tallinn Water official, interview with the author, 14 May 1998.

104. EBRD, "Lithuania: Kaunas Water and Environment Project," Memorandum to Board of Directors, BDS95-99, 30 June 1995.

105. The project's goals were to reduce heavy metals by 70 percent, total nitrogen by 10 percent, total phosphorus by 85 percent, BOD by 60 percent, and suspended solids by 33 percent. Ibid., Annex 5.

106. Author interviews with Linas Vainius, Lithuania Green Movement, 20 May 1998; an official from Kaunas Water Project Implementation Unit, 20 May, 1998; and an official from EBRD resident office in Riga, Latvia, 22 May, 1998.

107. Kaunas Water official, interview with the author, 20 May 1998.

108. The latter was also for projects in Cyprus, Malta, and Turkey.

109. The Bank lent money to the former Republic of Yugoslavia from the late 1970s until its break-up.

110. "Environmental Protection and EIB Finance," *Information* May 1982, 2–3.

111. EIB, *Annual Report* (Luxembourg: European Investment Bank, 1995), 44.

112. "The European Investment Bank: Accountable Only to the Market?" A Report by the CEE Bankwatch Network in cooperation with the Heinrich Böll Foundation, Brussels, December 1999, 7. An example of a project financed by the EIB after other MDBs turned it down for broader policy reasons is the new metro line in Budapest. The World Bank and Hungarian NGOs did not see the high costs of the metro project as justifying the benefits, and felt other potential projects, such as maintenance of existing roads, were greater priorities in Hungary's transport sector. Author interviews with World Bank official in Poland and Hungarian NGOs, October 1997. There are also examples of projects offered by the World Bank or EBRD that a recipient country had funded by the EIB instead, because there has been relatively less conditionality tied to EIB loans.

113. Peter Helger, "An Evaluation Study of Industrial Projects Financed by the European Investment Bank under the Objective of Regional Development" (Luxembourg: European Investment Bank, 1998), 2.

114. "The European Investment Bank: Accountable Only to the Market?" entire document. For evidence of prickly relations, see the exchange of correspondence between the Network and the Bank, Annex I.

115. Walter Hook, Deike Peters, Magdalena Stoczkeiwicz, and Wojciech Suchorzewski, "Transport Sector Investment Decision-Making in the Baltic Sea Region." Report by the Institute for Transportation and Development Policy for the Helsinki Commission Programme Implementation Task Force and Baltic 21, June 1999, vii. They note that nearly all World Bank road lending in CEE is for road maintenance or rehabilitation, as well as bypass routes around small towns, while the EBRD has financed both new highways and road maintenance.

116. CEE Bankwatch Network, "Blueprints for Sustainable Transportation in Central and Eastern Europe," May 1997.

117. This loan, for 95 million ECU, was signed in 1996. Its goal is to upgrade the M3 toll highway linking Budapest to Gyöngyös.

118. World Bank Transport Sector Project Completion Report No. 15445, cited in Walter Hook and Andras Lukacs, "The European Investment Bank's Loan to Hungary's M3 Highway: A Case Study," document from Hungary's Clean Air Action Group, 1997.

119. Ibid.

120. EIB project official, interview with the author, July 1996; Polish Finance Ministry official, interview with the author, October 1997.

Chapter Six: Conclusions

1. As noted in chapter 2, learning entails a cognitive evolution. Adaptation, in turn, involves behavioral change that does not entail a reexamination of underlying values or changes in the dynamics of decision making. See Ernst B. Haas, *When Knowledge Is Power: Three Models of Change in International Organizations* (Berkeley: University of California Press, 1990).

2. Also, related to this point, the banks do not explicitly address the impact of their portfolios on international accords such as the Kyoto Protocol on climate change. In the case of climate change, the banks are exploring joint implementation activities, but it remains to be seen how well such actions are integrated into their larger body of work.

3. Department for International Development, "Working in Partnership with the European Investment Bank," March 2000, 2.

4. Michelle Miller-Adams, *The World Bank: New Agendas in a Changing World* (New York: Routledge Studies in Development Economics, 1999), 77.

5. Rich, *Mortgaging the Earth,* 53.

6. International Financial Institution Advisory Commission (Meltzer Commission), http://phantom-x.gsia.cmu.edu/IFIAC/USMRPTDV.html.

7. Quoted in Gopinath, "Wolfensohn Agonistes," 39.

References

Books, Articles, and Papers

Ackermann, Richard. 1991. "Environment in East Central Europe: Despair or Hope." *Transition* 4: 9–11.

Adams, Lorraine. 1998. "World Bank Hires Auditors to Probe Its Own Spending." *Washington Post* (16 July): A1, 21.

Alcamo, Joseph, ed. 1992. *Coping with Crisis in Eastern Europe's Environment.* New York: Parthenon Publishing Group.

Allison, Graham. 1971. *Essence of Decision: Explaining the Cuban Missile Crisis.* Boston: Little, Brown.

Andrews, Richard N. L. 1993. "Environmental Policy in the Czech and Slovak Republic." In *Environment and Democratic Transition: Policy and Politics in Central and Eastern Europe,* edited by Anna Yari and Pal Tamas, 5–48. Boston: Kluwer Academic Publishers.

Ascher, William. 1983. "New Development Approaches and the Adaptability of International Agencies: The Case of the World Bank." *International Organization* 37(3): 415–439.

Aufderheide, Pat and Bruce Rich. 1988. "Environmental Reform and the Multilateral Development Banks." *World Policy Journal* 5: 301–321.

Barnes, James N. and Sandrine Bretonni. 1997. "An Overview of the European Investment Bank: Accountability and Transparency." Paris: Friends of the Earth.

Barnett, Michael N. and Martha Finnemore. 1999. "The Politics, Power, and Pathologies of International Organizations." International Organization 53(4): 699–732.

Baum, Warren C. 1982. *The Project Cycle.* Washington, D.C.: The World Bank.

Berg, Elliot and Don Sherk. 1994. "The World Bank and its Environmentalist Critics." In *Bretton Woods: Looking to the Future,* C 305–322. Washington, D.C.: Bretton Woods Commission.

Bochniarz, Zbigniew. 1990. Hearing before Subcommittee on Transport and Hazardous Materials of the Committee on Energy and Commerce (U.S. Environmental Initiatives in Eastern Europe), U.S. House of Representatives 101st Congress, 23 April, Serial No. 101–140.

Brandon, Katrina. 1994. "Environment and Development at the Bretton Woods Institutions." In *Bretton Woods: Looking to the Future,*C 133-142. Washington, D.C.: Bretton Woods Commission.

Brauchli, Marcus W. 1998. "World Bank Is Hurt by Its Failure to Anticipate the Indonesia Crisis." *Wall Street Journal,* 14 July: 1.

Brenton, Tony. 1994. *The Greening of Machiavelli.* London: The Royal Institute of International Affairs.

Bowley, Graham. 1996. "MDBs under Investigation." *Financial Times* (27 September): XXI.

Bretton Woods Commission. 1994. *Bretton Woods: Looking to the Future.* Washington, D.C.: The Bretton Woods Commission.

Brosshard, Peter, et al. 1996. *Lending Credibility: New Mandates and Partnerships for the World Bank.* Washington, D.C.: World Wildlife Fund.

Buchanan, James and Gordon Tullock. 1962. *The Calculus of Consent.* Ann Arbor: University of Michigan Press.

Bucknall, Julia. 2000. "Poland: Reducing the Costs of Complying with EU Environmental Legislation." In *Rural Development, Natural Resources, and the Environment,* ed. L. Alexander Norsworthy, 85–89. Washington, D.C.: The World Bank.

Cafruny, Alan W. and Glenda G. Rosenthal, eds. 1993. *The State of the European Community: The Maastricht Debates and Beyond.* Boulder: Lynne Rienner Publishers.

Caldwell, Lynton Keith. 1990. *International Environmental Policy: Emergence and Dimensions.* Durham, N.C.: Duke University Press.

Caulfield, Catherine. 1996. *Masters of Illusion: The World Bank and the Poverty of Nations.* New York: Henry Holt.

CEE Bankwatch Network. 1999. "The European Investment Bank: Accountable Only to the Market?" EU-Policy Paper no. 1. Brussels: Heinrich Böll Foundation.

CEE Bankwatch Network. 1997. "Blueprints for Sustainable Transportation in Central and Eastern Europe." May.

Cohen, Michael D., James G. March, et al. 1972. "A Garbage Can Model of Organizational Choice," *Administrative Science Quarterly* 17: 1–25.

Cole, G. D. H., and Margaret Cole. 1934. *A Guide to Modern Politics,* book 3. New York: Knopf.

Conable, Barber B. 1991. "To the World Resources Institute, Washington, D.C., May 5, 1987." In *The Conable Years at the World Bank: Major Policy Addresses of Barber B. Conable, 1986–91.* Washington, D.C.: The World Bank.

Connolly, Barbara. 1996. "Increments for the Earth: The Politics of Environmental Aid." In *Institutions for Environmental Aid: Pitfalls and Promise,* ed. Robert O. Keohane and Marc A. Levy, 327–365 (Cambridge: MIT Press).

Connolly, Barbara and Tamar Gutner. 2000. "Policy Networks and Process Diffusion: 'Environment for Europe.'" Paper prepared for delivery at the International Studies Association Annual Convention, Los Angeles.

Connolly, Barbara, and Tamar Gutner. 1997. "Organizational Inertia and Innovation: Environmental Aid to Central and Eastern Europe." Paper delivered at the 1997 Annual Meeting of the American Political Science Association, Washington, D.C.

Connolly, Barbara Tamar Gutner, and Hildegard Bedarff. 1996. "Organizational Inertia and Environmental Assistance to Eastern Europe." In *Institutions for Environmental Aid: Pitfalls and Promise,* ed. Robert O. Keohane and Marc A. Levy, 281–323. Cambridge: MIT Press.

Conradi, Peter. 1999. "EU Bank 'Lost £180m on the Bond Market,'" *Sunday Times,* 12 September, accessed online.

Covey, Jane G. 1998. "Is Critical Cooperation Possible? Influencing the World Bank Through Operational Collaboration and Policy Dialogue." In *The Struggle for Accountability: the World Bank, NGOs, and Grassroots Movements,* ed. Jonathan A. Fox and L. David Brown, 81–119. Cambridge: MIT Press.

Cox, Robert W. and Harold K. Jacobson. 1973. *The Anatomy of Influence: Decision Making in International Organization.* New Haven: Yale University Press.

Crockett, Tamara Raye and Cynthia B. Schultz. 1991. "The Integration of Environmental Policy and the European Community: Recent Problems of Implementation and Enforcement." *Columbia Journal of Transnational Law* 19(169): 181–182.

Danaher, Kevin, ed. 1994. *Fifty Years is Enough.* Boston: South End Press.

DeBardeleben, Joan, ed. 1991. *To Breathe Free: East Central Europe's Environmental Crisis.* Baltimore: Johns Hopkins Press.

DeSombre, Elizabeth R. and Joanne Kauffman. 1996. "The Montreal Protocol Multilateral Fund: Partial Success Story." In *Institutions for Environmental Aid: Pitfalls and Promise,* ed. Robert O. Keohane and Marc A. Levy, 89–126. Cambridge: MIT Press.

Downs, Anthony. 1957. *An Economic Theory of Democracy.* New York: Harper and Row.

EBRD. 1990. "Articles Establishing the European Bank for Reconstruction and Development." London: EBRD.

EBRD. 1992. *Annual Report.* London: EBRD.

EBRD. 1992. "Environmental Management: The Bank's Policy Approach." London: EBRD.

EBRD. 1993. *Annual Report.* London: EBRD.

EBRD. 1994. *Annual Report.* London: EBRD.

EBRD. 1995. *Annual Report.* London: EBRD.

EBRD. 1996. *Annual Report.* London: EBRD.

EBRD. 1997. *Annual Report.* London: EBRD.

EBRD. 1998. *Annual Report.* London: EBRD.

EBRD. 1999. *Annual Report.* London: EBRD.

Economist. 1999. "Clean up or Clear Out." *Economist* (11 December): 47.

Economist. 1990. "Cleaning up After Communism." *Economist* (17 February): 54.

EIB. 1989. *Annual Report.* Luxembourg: EIB.

EIB. 1990. *Annual Report.* Luxembourg: EIB.

EIB. 1991. *Annual Report.* Luxembourg: EIB.

EIB. 1992. *Annual Report.* Luxembourg: EIB.

EIB. 1993. *Annual Report.* Luxembourg: EIB.

EIB. 1994. *Annual Report.* Luxembourg: EIB.

EIB. 1995. *Annual Report.* Luxembourg: EIB.

EIB. 1996. *Annual Report.* Luxembourg: EIB.

EIB. 1997. *Annual Report.* Luxembourg: EIB.

EIB. 1998. *Annual Report.* Luxembourg: EIB.

EIB. 1999. *Annual Report.* Luxembourg: EIB.

European Commission. 2000. "The Weight of the Law." *Environment for Europeans,* March.

Fairman, David, and Michael Ross. 1996. "Old Fads, New Lessons: Learning from Economic Development Assistance." In *Institutions for Environmental Aid: Pitfalls and Promise,* ed. Robert O. Keohane and Marc A. Levy, 29–51. Cambridge: MIT Press.

Fidler, Stephen. 2001. "World Bank Employees Blame Wolfensohn for Morale Crisis." *Financial Times* (31 January).

Fisher, Duncan. 1992. "Paradise Deferred: Environmental Policymaking in Central and East Central Europe." London: Royal Institute of International Affairs and Ecological Studies Institute.

Fox, Jonathan A. and L. David Brown. 1998. *The Struggle for Accountability: The World Bank, NGOs, and Grassroots Movements.* Cambridge: MIT Press.

French, Hilary F. 1992. "Eastern Europe's Clean Break with the Past." Worldwatch, March/April.

Galambos, Judit. 1992. "Political Aspects of an Environmental Conflict: The Case of the Gabcikovo-Nagymaros Dam System." In *Perspectives on Environmental Conflict and International Relations,* ed. Jyrki Kakonen, 72–95. London: Printer Publishers.

George, Susan and Fabrizio Sabelli. 1994. *Faith and Credit: The World Bank's Secular Empire.* Boulder: Westview Press.

Gerster, Richard. 1993. "Accountability of Executive Directors in the Bretton Woods Institutions." *Journal of World Trade* 27(6): 87–116.

Goldberg, Donald M., et al. 1995. "The European Bank for Reconstruction and Development: An Environmental Progress Report." Washington, D.C.: CIEL.

Gopinath, Deepak. 2000. "Wolfensohn Agonistes." *Institutional Investor* (September): 33–48.

Gourevitch, Peter Alexis. 1996. "Squaring the Circle: The Domestic Sources of International Cooperation," *International Organization* 50(2): 349–373.

Graham, George. 1995. "IFC to Leave Privatization to Banks." *Financial Times* (8 December): 4.

———. 1989. "France to Press Proposals for European Development Bank." *Financial Times,* (1 December): 4.

Greenpeace. 1994. "World Bankenstein: A Monstrous Institution." *Greenpeace* 3(4) (October–December): 2.

Grieco, Joseph. 1988. "Anarchy and the Limits of Cooperation: A Realist Critique of the Newest Liberal Institutionalism." *International Organization* 42(3): 485–507.

Grubb, Michael, et al. 1993. *The "Earth Summit" Agreements: A Guide and Assessment.* London: Earthscan Publications.

Gutner, Tamar and Stacy D. VanDeveer. 2001. "The Role and Impacts of Mature Policy Networks in the Baltic Sea Region." Paper delivered at International Studies Association conference, Chicago, IL.

Gwin, Catherine. 1994. *U.S. Relations with the World Bank.* Washington, D.C.: The Brookings Institution.

Haggard, Stephan and Andrew Moravcsik. 1993. "The Political Economy of Financial Assistance to Eastern Europe, 1989–1991." In *After the Cold War: International Institutions and State Strategies in Europe, 1989–1991,* ed. Robert O. Keohane, Joseph S. Nye, and Stanley Hoffmann, 246–285. Cambridge: Harvard University Press.

Haas, Ernst B. 1990. *When Knowledge Is Power: Three Models of Change in International Organizations.* Berkeley: University of California Press.

Haas, Peter M. and Ernst B. Haas. 1995. "Learning to Learn: Improving International Governance." *Global Governance* 1(September–December): 255–285.

Haas, Peter M. 1992. "Introduction: Epistemic Communities and International Policy Coordination." *International Organization* 46(1): 1–36.

Hall, Peter A. and Rosemary C. R. Taylor. 1996. "Political Science and the Three Institutionalisms." *Political Studies* 44: 952–973.

Handyside, Gillian. 1997. "EU Hopefuls Face $130 Billion Environmental Bill." *Reuters* (10 September).

Helger, Peter. 1998. "An Evaluation Study of Industrial Projects Financed by the European Investment Bank under the Objective of Regional Development." Luxembourg: European Investment Bank.

Hertzman, Clyde. 1995. "Environment and Health in Central and Eastern Europe: A Report for the Environmental Action Programme for Central and Eastern Europe." Washington, D.C.: The World Bank.

Hirschmann, Albert O. 1967. *Development Projects Observed.* Washington, D.C.: The Brookings Institution.

Horberry, John. 1985. "The Accountability of Development Assistance Agencies: The Case of Environmental Policy." *Ecology Law Quarterly* 12: 817–869.

Hörhager, Axel. 1992. "An Environmental Perspective for Eastern Europe." *EIB Papers* 18 (November): 35–57.

Hughes, Gordon. 1995. "Is the Environment Getting Cleaner in Central and Eastern Europe: Selected Evidence for Air Pollution and Drinking Water Contamination." Washington, D.C.: World Bank.

Hunter, David. 1996. "The Role of the World Bank in Strengthening Governance, Civil Society, and Human Rights." In *Lending Credibility: New Mandates and Partnerships for the World Bank,* ed. Peter Brossard et al., 65–80. Washington, D.C.: World Wildlife Fund.

Institute for International Finance. 1999. "Capital Flows to Emerging Market Economies: Update." Washington, D.C.: Institute for International Finance (25 September).

Jancar, Barbara. 1990. "Democracy and the Environment in Eastern Europe and the Soviet Union," *Harvard International Review* 12(4): 13–18, 58–59.

Kabala, Stanley J. 1991. "Environment and Development in the New East Central Europe: Addressing the Environmental Legacy of Central Planning." Middlebury, VT: Geonomics Institute.

Kapur, Devesh, John P. Lewis, and Richard Webb. 1997. *The World Bank: Its First Half Century.* 2 vols. Washington, D.C.: The Brookings Institution.

Keck, Margaret E. 1998. "Planafloro in Rondônia: The Limits of Leverage." In *The Struggle for Accountability: The World Bank, NGOs, and Grassroots Movements,* ed. Jonathan A. Fox and L. David Brown, 181–218. Cambridge: MIT Press.

Kemple, Penn. 1989. "East Europe: Greening of the Reds." *Washington Post* (24 December): C3.

Keohane, Robert O. 1984. *After Hegemony: Cooperation and Discord in the World Political Economy.* Princeton: Princeton University Press.

Keohane, Robert O. and Marc A. Levy, eds. 1996. *Institutions for Environmental Aid: Pitfalls and Promise.* Cambridge: MIT Press.

Kiewiet, D. Roderick, and Mathew D. McCubbins. 1991. *The Logic of Delegation: Congressional Parties and the Appropriations Process.* Chicago: University of Chicago Press.

Kingdon, John W. 1984. *Agendas, Alternatives, and Public Policies.* Boston: Little, Brown.

Klarer, Jürg, and Bedrich Moldan, eds. 1997. *The Environmental Challenge for Central European Countries in Transition.* New York: John Wiley.

Korten, Frances F. 1993. "The High Costs of Environmental Loans." *Asia Pacific Issues: Analysis from the East-West Center,* 9: 1–8.

Kraske, Jochen et al. 1996. *Bankers with a Mission: The Presidents of the World Bank, 1946–91.* Washington, D.C.: Oxford University Press.

Johnson, Stanley P. and Guy Corcelle. 1995. *The Environmental Policy of the European Communities.* Boston: Kluwer Law International.

Lambert, Richard and David Marsh. 1994. "Credit Where It Is Due: The New President of the EBRD Has Brought Prudence Rather than Vision to the Bank." *Financial Times* (29 December): 9.

Landau, Luis and Julius Gwyer. 1997. *Poland Country Assistance Review: Partnership in a Transition Economy*. Washington, D.C.: The World Bank.

Lankowski, Carl. 1984. "Environmental Impact Review in the European Investment Bank." Washington, D.C.: School of International Service, American University.

Le Prestre, Philippe G. 1989. *The World Bank and the Environmental Challenge*. Cranbury, N.J.: Associated University Presses.

Lewenhak, Sheila. 1982. *The Role of the European Investment Bank*. London: Croom Helm.

Liberatore, Angela. 1991. "Problems of Transnational Policymaking: Environmental Policy in the European Community." *European Journal of Political Research* 19: 281–305

Lodge, Juliet, ed. 1993. *The European Community and the Challenge of the Future*. New York: St. Martin's Press.

March, James G. and Herbert A. Simon. 1958. *Organizations*. New York: John Wiley.

March, James G. and Johan P. Olsen. 1989. *Rediscovering Institutions: The Organizational Basis of Politics*. New York: The Free Press.

March, James G. 1978. "Bounded Rationality, Ambiguity, and the Engineering of Choice." *Bell Journal of Economics* 9(2): 587–608.

Martin, Lisa L. and Beth A. Simmons. 1998. "Theories and Empirical Studies of International Institutions." *International Organization* 52(4): 729–757.

Martinson, Jane. 1995. "EBRD Criticized on N-plant Loan." *Financial Times* (15 February): 2.

Mason, Edward S. and Robert E. Asher. 1973. *The World Bank Since Bretton Woods*. Washington, D.C.: The Brookings Institution.

Mattli, Walter and Anne-Marie Slaughter. 1998. "Revisiting the European Court of Justice," *International Organization* 52(1): 177–209.

McNamara, Robert. 1981. *The McNamara Years at the World Bank: Major Policy Address of Robert S. McNamara, 1968–1971*. Baltimore: Johns Hopkins University Press.

Mearsheimer, John. 1994/95. "The False Promise of International Institutions." *International Security* (winter): 5–49.

Miller-Adams, Michelle. 1999. *The World Bank: New Agendas in a Changing World*. New York: Routledge.

Moldan, Bedrich and Jerald L. Schnoor. 1992. "Czechoslovakia: Examining a Critically Ill Environment." *Environment Science and Technology* 26(1): 16.

Moravcsik, Andrew. 1998. *The Choice for Europe: Social Purpose and State Power from Messina to Maastricht*. Ithaca: Cornell University Press.

Morse, Bradford and Thomas Berger. 1992. *Sardar Sarovar: Report of the Independent Review.* Ottawa, Canada: Resources Future International.

Mosley, Paul, Jane Harrigan, and John Toye. 1991. *Aid and Power: The World Bank and Policy-based Lending,* vol. 1. New York: Routledge.

Naim, Moises. 1994. "The World Bank: Its Role, Governance and Organizational Culture." In *Bretton Woods: Looking to the Future,* C 273–287. Washington, D.C.: Bretton Woods Commission.

Nelson, Joan M. and Stephanie J. Eglinton. 1993. *Global Goals, Contentious Means: Issues of Multiple Aid Conditionality.* Washington, D.C.: Overseas Development Council.

Nelson, Paul J. 1995. *The World Bank and Non-Governmental Organizations.* New York: St. Martin's Press.

"NGOs Criticise EBRD Lending, Environment Record." 2001. *Reuters News Service,* 23 April.

OECD. 1996. *Environmental Indicators: A Review of Selected Central and Eastern European Countries.* Paris: OECD.

Operations Evaluation Department. 1994. *1992 Evaluation Results.* Washington, D.C.: The World Bank.

Owen, Henry. 1994. "The World Bank: Is 50 Years Enough?" *Foreign Affairs* 73(5): 97–108.

Payer, Cheryl. 1982. *The World Bank: A Critical Analysis.* New York: Monthly Review Press.

Pearce, David W. and Jeremy J. Warford. 1993. *World without End: Economics, Environment, and Sustainable Development.* New York: Oxford University Press.

PHARE. 1998. *Annual Report.* Brussels: European Commission.

Philips, Michael. 1991. *The Least-Cost Energy Path for Developing Countries.* Washington, D.C.: International Institute for Energy Conservation.

Pierson, Paul. 1996. "The Path to European Integration." *Comparative Political Studies* 29(2): 123–163.

Pollack, Mark A. 1997. "Delegation, Agency, and Agenda Setting in the European Community." *International Organization* 51(1): 99–134.

Powell, Walter W. and Paul J. DiMaggio, eds. 1991. *The New Institutionalism in Organizational Analysis.* Chicago: University of Chicago Press.

Preston, Robert and Jimmy Burns. 1993. "EBRD Spends More on Itself than It Hands Out in Loans." *Financial Times* (13 April): 1.

Ranis, Gustav. 1994. "Defining the Mission of the World Bank Group." In *Bretton Woods: Looking to the Future,* C 73-82. Washington. D.C.: Bretton Woods Commission.

Reed, David, ed. 1992. *Structural Adjustment and the Environment.* Boulder: Westview Press.

Rich, Bruce. 1994. *Mortgaging the Earth: The World Bank, Environmental Impoverishment, and the Crisis of Development.* Boston: Beacon Press.

Rich, Bruce. 1997. "Epilogue: The Gorbachav of the World Bank?" Environmental Defense Fund. Photocopy.

Ross, Michael. 1996. "Conditionality and Logging Reform in the Tropics." In *Institutions for Environmental Aid: Pitfalls and Promise,* ed. Robert O. Keohane and Marc A. Levy, 167–197. Cambridge: MIT Press.

Rotberg, Eugene H. 1994. "Financial Operations of the World Bank." In *Bretton Woods: Looking to the Future,* C 185–214. Washington, D.C.: Bretton Woods Commission.

Ruggie, John Gerard. 1993. "Multilateralism: The Anatomy of an Institution." In *Multilateralism Matters,* ed. John Gerard Ruggie, 3–47. New York: Columbia University Press.

Sabatier, Paul A. 1988. "An Advocacy Coalition Framework of Policy Change and the Role of Policy-Oriented Learning Therein." *Policy Sciences* 21: 131.

Sanford, Jonathan Earl. 1988. "U.S. Policy Toward the Multilateral Development Banks: The Role of Congress." *The George Washington Journal of International Law and Economics* 22(1): 1–115.

Sbragia, Alberta M., ed. 1992. *Euro-Politics: Institutions and Policymaking in the "New" European Community.* Washington, D.C.: The Brookings Institution.

———. 1993. "EC Environmental Policy: Atypical Ambitions and Typical Problems?" In *The State of the European Community: The Maastricht Debates and Beyond,* ed. Alan W. Cafruny and Glenda G. Rosenthal, 337–352. Boulder: Lynne Rienner Publishers.

Schoultz, Lars. 1982. "Politics, Economics, and U.S. Participation in Multilateral Development Banks." *International Organization* 36(3): 537–574.

Schreiber, Helmut. 1990. "The Threat From Environmental Destruction in Eastern Europe," *Journal of International Affairs* 44 (winter): 359–391.

Seymour, Francis. 1996. "Overview." In *Lending Credibility: New Mandates and Partnerships for the World Bank,* ed. Peter Brossard et al., 1–28. Washington, D.C.: World Wildlife Fund.

Shepsle, Kenneth A. 1989. "Studying Institutions: Some Lessons from the Rational Choice Approach." *Journal of Theoretical Politics* 1(2): 131–147.

Shihata, Ibrahim F. I. 1990. *The European Bank for Reconstruction and Development: A Comparative Analysis of the Constituent Agreement.* London: Graham and Trotman/Martin Nijhoff.

———. 1994. *The World Bank Inspection Panel.* New York: Oxford University Press.

———. 1995. *The World Bank in a Changing World: Selected Essays and Lectures,* vol. II. Boston: Martinus Nijhoff.

Steinmo, Sven, Kathleen Thelen, and Frank Longstreth. 1992. *Structuring Politics: Historical Institutionalism in Comparative Analysis.* New York: Cambridge University Press.

Stern, Nicholas, and Francisco Ferreira. 1997. "The World Bank as 'Intellectual Actor.'" In *The World Bank: Its First Half Century,* ed. Devesh Kapur, John P.

Lewis, and Richard Webb, 523–610. Washington, D.C.: The Brookings Institution.

Strange, Susan. 1983. "Cave! hic dragones: A Critique of Regime Analysis." In *International Regimes,* ed. Stephen Krasne, 337–354. Ithaca: Cornell University Press.

Taylor, Serge. 1984. *Making Bureaucracies Think: The Environmental Impact Statement Strategy of Administrative Reform.* Stanford: University of California Press.

Tendler, Judith. 1975. *Inside Foreign Aid.* Baltimore: Johns Hopkins University Press.

Thatcher, Peter S. 1992. "The Role of the United Nations." In *The International Politics of the Environment,* ed. Andrew Hurrell and Benedict Kingsbury, 183–211. New York: Oxford University Press.

Tucker, Emma. 1995. "Bank Prepares to Bite Nuclear Bullet." *Financial Times* (12 April): 2.

Udall, Lori. 1998. "The World Bank and Public Accountability: Has Anything Changed?" In *The Struggle for Accountability: The World Bank, NGOs, and Grassroots Movements,* ed. Jonathan A. Fox and L. David Brown, 391–436. Cambridge: MIT Press.

Wade, Robert. 1997. "Greening the Bank: The Struggle over the Environment, 1970–95." In *The World Bank: Its First Half Century,* vol. 2, ed. Devesh Kapur, John P. Lewis, and Richard Webb, 611–736. Washington, D.C.: The Brookings Institution.

Wade, Robert. 2001. "Showdown at the World Bank." *New Left Review* 7: 124–137.

Wapenhans, Willi A. 1994. "Efficiency and Effectiveness: Is the World Bank Group Well Prepared for the Task Ahead?" In *Bretton Woods: Looking to the Future,* C 289-304. Washington, D.C.: Bretton Woods Commission.

Ward, Barbara and Rene Dubos. 1972. *Only One Earth: The Care and Maintenance of a Small Planet.* New York: Norton.

Weber, Steven. 1994. "Origins of the European Bank for Reconstruction and Development." *International Organization* 48(1): 1–38.

Wenning, Marianne. 1992. "Greening the European Investment Bank." WWF International Discussion Paper (December).

Weir, Margaret and Theda Skocpol. 1985. "State Structures and the Possibilities for 'Keynesian' Responses to the Great Depression in Sweden, Britain and the United States." In *Bringing the State Back In,* ed. Peter Evans, Dietrich Rueschemeyer, and Theda Skocpol, 107–163. Cambridge: Cambridge University Press.

Wendt, Alexander. 1992. "Anarchy Is What States Make of It: The Social Construction of Power Politics." *International Organization* 46(2): 391–425.

Williamson, Oliver. 1981. "The Economics of Organization: The Transaction Cost Approach." *American Journal of Sociology* 87(3): 548–577.

Wilson, James Q. 1989. *Bureaucracy: What Government Agencies Do and Why They Do It.* New York: Basic Books.

Wold, Chris A. and Durwood Zaelke. 1992. "Promoting Sustainable Development and Democracy in Central and Eastern Europe: The Role of the European Bank for Reconstruction and Development." *American University Journal of International Law and Policy* 7(3): 559–564.

World Bank. 1990. *Annual Report.* Washington, D.C.: World Bank.

World Bank. 1991. *Annual Report.* Washington, D.C.: World Bank.

World Bank. 1992. *Annual Report.* Washington, D.C.: World Bank.

World Bank. 1992. *World Development Report.* Washington, D.C.: World Bank.

World Bank. 1993. *Annual Report.* Washington, D.C.: World Bank.

World Bank. 1994. *Annual Report.* Washington, D.C.: World Bank.

World Bank. 1994. *Making Development Sustainable* . Washington, D.C.: World Bank.

World Bank. 1995. *Annual Report.* Washington, D.C.: World Bank.

World Bank. 1995. *Mainstreaming the Environment.* Washington, D.C.: World Bank.

World Bank. 1995. *The World Bank and the Environment in Central and Eastern Europe: 1990–95.* Washington, D.C.: World Bank.

World Bank. 1996. *Annual Report.* Washington, D.C.: World Bank.

World Bank. 1997. *Annual Report.* Washington, D.C.: World Bank.

World Bank. 1997. *Global Development Finance.* Washington, D.C.: World Bank.

World Bank. 1997. *Private Capital Flows to Developing Countries: The Road to Financial Integration.* New York: Oxford University Press.

World Bank. 1998. *Annual Report.* Washington, D.C.: World Bank.

World Bank. 1999. *Annual Report.* Washington, D.C.: World Bank.

World Bank. 2000. "Promoting Environmental Sustainability in Development: An Evaluation of the World Bank's performance," Operations Evaluation Department, Sector and Thematic Evaluation Group, September 15.

World Bank. 2000. "Toward an Environmental Policy for the World Bank Group." Progress Report/Discussion Draft, April.

World Commission on Environment and Development. 1987. *Our Common Future.* New York: Oxford University Press.

Wurzel, Rüdiger. 1993. "Environmental Policy." In *The European Community and the Challenge of the Future,* ed. Juliet Lodge, 178–199. New York: St. Martin's Press.

Yasutomo, Dennis T. 1995. *The New Multilateralism in Japan's Foreign Policy.* New York: St. Martin's Press.

Young, Oran. 1994. *International Governance: Protecting the Environment in a Stateless Society.* Ithaca: Cornell University Press.

Zylicz, Tomasz. 1994. "In Poland, It's Time for Economics." *Environmental Impact Assessment Review* 14(2,3): 79–94.

Other Documents

Bank Information Center. February 1998. "Tuesday Group Report."

Bank Information Center. April 1999. "Tuesday Group Report." Volume 3, no. 4.

Brady, Nicholas, letter to William K. Reilly, 22 February 1990.

"The CEE Bankwatch Network for Monitoring International Financial Institutions." 1997. Krakow, March 4.

Christie, Herbert. June 1992. "The European Investment Bank: Finance for Environmental Protection in Europe." Speech presented at the Earth Summit/United Nations Conference on Environment and Development, Rio de Janeiro.

Department for International Development. March 2000. "Working in Partnership with the European Investment Bank," London.

Development Committee Task Force on Multilateral Development Banks. 1996. "Serving a Changing World." Washington, D.C.: March 15.

"E-note." August 1997. International Institute for Energy Conservation.

"East European Energy Wastage to be Tackled by EBRD and Landis and Gyr Partnership," EBRD Press Release, London (18 December 1996).

EBRD. 1997. "Environments in Transition: the Environmental Bulletin of the EBRD." London.

EBRD. 1996. "Revised Environmental Procedures." BDS96-23, London.

EBRD. June 1994. "Estonia: Tallinn Water and Environment Project." Report to Board of Directors, BDS94-93.

EBRD. 1993. "Environments in Transition: The Environmental Bulletin of the EBRD." London.

EBRD. November 1993. "Former Yugoslav Republic of Macedonia: Power Sub-Sector Project." Report to Board of Directors, BDS93-157.

EBRD. November 1992. "Estonia Energy Sector Emergency Investment Project." Report to Board of Directors, BDS92-124.

EBRD. "EBRD Activities in Poland," http://www.ebrd.com/english/opera/country/polafact.htm.

EBRD. "EBRD Activities in Hungary," http://www.ebrd.com/english/opera/country/hungact.htm.

EBRD. 21 March 1997. "Independent Safety Panel's Recommendations on Ignalina NPP Released by Nuclear Safety Account," http://www.ebrd.com/english/opera/nucsafe/prelease/18mar21.htm.

"Ecological Problems of Siauliai," Siauliai Municipality Environmental Affairs Department, 1994.

Effective Implementation: Key to Development Impact. Portfolio Management Task Force. Confidential Report of the World Bank, September 22, 1991.

EIB. 1995. "C'EZ Power Plant Improvement." Report to the Board of Directors.

EIB. 1996. "Environmental Policy Statement."

EIB. 4 June 1984. "Board of Governors: Bank Activity," Report by the Board of Directors, Document 84/4.

EIB. May 1982. *Information*. No. 30.

EIB. Second Quarter 1996. *Information*. No. 88.

EIB. "Rules on Public Access to Documents." Adopted by the Bank's Management Committee on 26 March 1997. 97/C 243/06.

EIB. "Environmental Protection and EIB Finance," *Information* May 1982.

"Environmental Policy in Central and Eastern Europe," World Bank document, EMTEN, 19 November 1991.

Environmental Defense Fund. "World Bank Plans Increased Support for Timber Harvesting," May 1994, accessed at http://222edf.org/pubs/newsreleases/1994/may/e%Fwforestr.html.

European Communities, Commission of the. "Working Procedures between the EIB and the Commission Services (DGXI and XXII) in the Consultation of the Commission under Article 21 of the EIB Statute," Document. 10/30/92.

European Communities, Commission of the. 1991. "Report on EIB Operations Outside the Community."

Garg, Prem. May 13, 1997. Office Memorandum, "Portfolio Monitoring Report."

Green Horizon, vol. 2, no. 8, October 1999.

Gutner, Tamar. December 1990. "Bankrolling Eastern Europe." *Prodigy News Service.*

Hook, Walter and Andras Lukacs. 1997. "The European Investment Bank's Loan to Hungary's M3 Highway: A Case Study," Document from Hungary's Clean Air Action Group.

Hook, Walter, Deike Peters, Magdalena Stoczkeiwicz, and Wojciech Suchorzewski. 1999. "Transport Sector Investment Decision-making in the Baltic Sea Region." Report by the Institute for Transportation and Development Policy for the Helsinki Commission Programme Implementation Task Force and Baltic 21.

"Implementing Community Environmental Law." 1996. Communication to the Council of the European Union and the European Parliament.

The International Financial Institution Advisory Commission (Meltzer Commission) report. March 2000. Available at http://phantom-x.gsia.cmu.edu/IFIAC/USMRPTDV.html.

"Is Nuclear Safety Account Going to Fail?" March 1998. Press Release, CEE Bankwatch.

Jordan, Lisa. 1997. "Sustainable Rhetoric vs. Sustainable Development: The Retreat from Sustainability in World Bank Development Policy." Bank Information Center.

"K2/R4 Completion Projects," http://www.ecn.cz/k2r4/BOARD.sTM#2.2.

Klaus, Vaclav, "Remarks of the Czech Prime Minister Vaclav Klaus to the Bretton Woods Committee," Washington, D.C., Friday, 15 October 1993.

"METAP: Helping to Preserve and Manage a Shared Resource in the Mediterranean," *Information* 88, second quarter (1996): 11.

OECD, "Report on Environmental Financing in CEEC/NIS," Task Force for the Implementation of the Environmental Action Programme in Central and Eastern Europe, document CCNM/ENV/EAP (98)24/REV2, 18 May 1998.

OECD. "Aid at a Glance," http://www.oecd.org/dac/htm/agdeu.htm.

Öko-Institute. 1995. "Statement Concerning the Least Cost Study for the Public Participation Programme Related to the Project 'Completion of the Mochovce NPP (Slovak Republic).'" Freiburg, Germany: Institute für Angewandte Ökologie.

"PPC Report to the Third Ministerial Conference 'Environment for Europe' in Sofia." 1995. London: Project Preparation Committee Secretariat.

PHARE. 1991. "Environment Sector Strategy for Central and Eastern Europe." PHARE internal strategy paper.

"Power Failure: A Review of the World Bank's Implementation of Its New Energy Policy." March 1994, Environmental Defense Fund, Natural Resources Defense Council.

Rich, Bruce and Stephanie Fried. Letter to James Wolfensohn from Environmental Defense Fund, 6 August 1998.

"SEED Act Implementation Report," Fiscal Year 1997, U.S. Department of State, February 1998.

USAID Congressional Presentations, FY 1990–2001.

U.S. Senate. Statement by the Honorable David C. Mulford, Under Secretary of Treasury for International Affairs before the Senate Foreign Relations Committee, Subcommittee on International Economic Policy, Trade, Oceans and the Environment, 22 March 1990.

U.S. Senate. Subcommittee on International Economic Policy, Trade, Oceans and Environment, Committee on Foreign Relations. *Statement of Bruce M. Rich on Behalf of Environmental Defense Fund, Friends of the Earth, National Audubon Society, National Wildlife Federation, Sierra Club Concerning Public International Financial Institutions: Environmental Performance and Management.* 3 March 1994.

"Vademecum: Procedures for the Commission's Opinion on EIB operations." 1996. European Commission, DGII.

Wapenhans, Willi A. 1991. *Effective Implementation: Key to Development Impact.* Portfolio Management Task Force. Confidential Report of the World Bank, 22 September.

Wenning, Marianne. 1992. "Greening the European Investment Bank." Switzerland: WWF International.

World Bank. 1990. *Staff Appraisal Report: Poland Energy Resource Development Project.*

World Bank. 1992. "Central Europe Department Projects Related to Energy/Environment." EMNTEN document, August/September.

World Bank. 1992. *Staff Appraisal Report: Czech and Slovak Republic Power and Environmental Improvement Project.*

World Bank. 1994. *Environmental Action Programme for Central and Eastern Europe.* Document submitted to the Ministerial Conference in Lucerne Switzerland (28–30 April 1993). Washington, D.C.: The World Bank.

World Bank. 1994. *Staff Appraisal Report: Estonia District Heating Rehabilitation Project.*

World Bank. 1994. *Staff Appraisal Report: Republic of Lithuania Klaipeda Environment Project.*

World Bank. 1995. *Staff Appraisal Report: Republic of Estonia: Haapsalu and Matsalu Bays Environment Project.*

World Bank. 1995. *Staff Appraisal Report: Republic of Lithuania Siauliai Environment Project.*

World Bank. 1996. "Effectiveness of Environmental Assessments and National Environmental Action Plans." 28 June, Operations Evaluation Department Report 15835.

World Bank. 1996. "Environment Matters: Annual Review." Washington, D.C.: The World Bank.

World Bank. 1997. *Implementation Completion Report: Poland Environment Management Project.* Washington, D.C.: World Bank.

World Bank. 1998. "Memorandum of the President of IBRD and IDA to the Executive Directors on a Country Assistance Strategy of the World Bank Group for Hungary."

World Bank. 2001. "Making Sustainable Commitments: An Environment Strategy for the World Bank."

World Bank. 2001. "The Strategic Compact: A Summary Note." March 13. http://www.worldbank.org.

World Bank Quality Assurance Group. 1997. "Portfolio Improvement Program: Draft Reviews of Sector Portfolio and Lending Instruments: A Synthesis." 24 April.

"The World Bank Streamlines Its Strategy for Transition Economies." February 1997. Transition, World Bank.

Interviews

"Off-the-record" interviews include:

EBRD officials from Executive Directors' offices, Municipal and Environmental Infrastructure, Energy Efficiency Unit, Environmental Appraisal Unit, various country teams, and representatives in Hungary, Latvia, Poland: October 1994; February–April 1995; June–August 1996; January 1997; May 1998.

EIB officials from the Projects Directorate, country units, Management Committee, Legal Directorate: April–May 1995; June–July 1996; November 1997.

European Commission, PHARE officials: August 1996, May 1998.

OECD official, July 1997.

Project Preparation Committee officials: in Washington, D.C., October, December 1995, July 1997; in London, March 1995, July 1996.

U.K. Treasury official, July 1996.

U.S. Treasury officials: October, November 1994, February, March, July, August 1995; March 1996.

U.S. AID official, Environmental and Natural Resources Division, Bureau for Europe/New Independent States, July 1997.

U.S. EPA officials, July, September 1995.

World Bank officials from EMTEN, QAG, Executive Directors' offices and units designing projects in CEE, and representatives in Hungary, Lithuania, Poland: October-February 1995; August 1995; January, March, May 1996; April, July and August 1997; March, May 1998.

In the Czech Republic, July 1992, interviews included officials from: Czechoslovakia Federal Committee for the Environment, Ministry of the Environment, Institute of Geography.

In Estonia, May–June 1998, interviews included officials from: Tallinn Waste and Sewerage Municipal Enterprise, Ministry of Environment, Haapsalu Water Works.

In Hungary, June, July 1992, June 1993, October-November 1997, interviews included officials from: Central Bank, Ministry of Environment, Ministry of Finance, Federal Committee for the Environment, National Authority for the Environment, Municipality of the City of Budapest, parliamentary committee for environmental protection.

In Latvia, May–June 1998, interviews included officials from: Liepaja Environment Project, Ministry of Environment.

In Lithuania, May–June 1998, interviews included officials from: Klaipeda Water Company, Siauliai Vendenys, Siauliai municipality, UAB Geoterma, Ministry of Environment, Kaunas Water.

In Poland, October–November 1997, interviews included officials from: National Environmental Fund, Ministry of Environmental Protection, Ministry of Finance.

Other interviews:

Ada Amon, October 1997.

Ernst-Günther Bröder, March 1997.

Kerstin Canby, March 1998.

József Feiler, October 1997.

Judit Galambos, June 1992.

Alex Hittle, November 1992.

Stephen Lintner, March 1998.
Andras Lukacs, October 1997.
Peep Mardiste, May 1998.
Bruce Rich, October 1994.
Wojciech Stolduski, October 1997.
Magda Stoczkiewicz, October 1997.
Jernej Stritih, October 1997.
Tomasz Terecki, October 1997.
Magda Toth Nagy, July 1992, October 1997.
Linas Vainius, May 1998.
János Vargha, July 1992.
Durwood Zaelke, August 1995.

Index